From Qubits to the Unified SuperStandard Model With Embedded SuperStrings

Stephen Blaha Ph. D.
Blaha Research

Pingree-Hill Publishing

ISBN: 978-0-9984575-9-8

Rev. 00/00/01 July 27, 2017

To Margaret

Some Other Books by Stephen Blaha

All the Megaverse! Starships Exploring the Endless Universes of the Cosmos using the Baryonic Force (Blaha Research, Auburn, NH, 2014)

SuperCivilizations: Civilizations as Superorganisms (McMann-Fisher Publishing, Auburn, NH, 2010)

PHYSICS IS LOGIC PAINTED ON THE VOID: Origin of Bare Masses and The Standard Model in Logic, U(4) Origin of the Generations, Normal and Dark Baryonic Forces, Dark Matter, Dark Energy, The Big Bang, Complex General Relativity, A Megaverse of Universe Particles (Blaha Research, Auburn, NH, 2015).

The Origin of Higgs ("God") Particles and the Higgs Mechanism: Physics is Logic III, Beyond Higgs – A Revamped Theory With a Local Arrow of Time, The Theory of Everything Enhanced, Why Inertial Frames are Special, Universes of the Mind (Blaha Research, Auburn, NH, 2015).

New Types of Dark Matter, Big Bang Equipartition, and A New U(4) Symmetry in the Theory of Everything: Equipartition Principle for Fermions, Matter is 83.33% Dark, Penetrating the Veil of the Big Bang, Explicit QFT Quark Confinement and Charmonium, Physics is Logic V (Blaha Research, Auburn, NH, 2015).

New Boson Quantum Field Theory, Dark Matter Dynamics, Dark Matter Fermion Layer Mixing, Genesis of Higgs Particles, New Layer Higgs Masses, Higgs Coupling Constants, Non-Abelian Higgs Gauge Fields, Physics is Logic VII (Blaha Research, Auburn, NH, 2015)

CQMechanics: A Unification of Quantum & Classical Mechanics, Quantum/Semi-Classical Entanglement, Quantum/Classical Path Integrals, Quantum/Classical Chaos (Blaha Research, Auburn, NH, 2016).

All the Universe! Faster Than Light Tachyon Quark Starships & Particle Accelerators with the LHC as a Prototype Starship Drive Scientific Edition (Pingree-Hill Publishing, Auburn, NH, 2011).

From Asynchronous Logic to The Standard Model to Superflight to the Stars; Volume 2: Superluminal CP and CPT, U(4) Complex General Relativity and The Standard Model, Complex Vierbein General Relativity, Kinetic Theory, Thermodynamics (Blaha Research, Auburn, NH, 2012)

New Boson Quantum Field Theory, Dark Matter Dynamics, Dark Matter Fermion Layer Mixing, Genesis of Higgs Particles, New Layer Higgs Masses, Higgs Coupling Constants, Non-Abelian Higgs Gauge Fields, Physics is Logic VII (Blaha Research, Auburn, NH, 2015)

The Origin of Fermions and Bosons,and Their Unification (Pingree-Hill Publishing, Auburn, NH, 2017).

Megaverse: The Universe of Universes (Pingree Hill Publishing, Auburn, NH, 2017).

SuperSymmetry and the Unified SuperStandard Model (Pingree Hill Publishing, Auburn, NH, 2017).

Available on Amazon.com, bn.com Amazon.co.uk and other international web sites as well as at better bookstores (through Ingram Distributors).

CONTENTS

FIGURES and TABLES

INTRODUCTION

This book provides an explanation of the number of space-time dimensions in our universe based on the number of qubit interactions and on the nature of the Asynchronous Logic parallelism requirements (both discussed in earlier books). Both approaches set the number of dimensions of our space-time to 4.

The book also shows that our Unified SuperStandard Model can be viewed as a distinct variant of SuperString theory due to the presence of strings within dressed free particles and the presence of SuperSymmetry. In brief the theory dresses particles with strings unlike SuperString theories which make strings into particles.

These topics are followed by a close study of the nature of a derivation of a fundamental theory of physics based on a three player model: Unchanging Concept of Reality (Theory), Map to material Reality, and the 'Unmoved Mover.'

It finishes with a detailed outline of the derivation of the Unified SuperStandard Model. Some aspects of the model are changed and typos in previous books hopefully corrected.

1. The Genesis of 4-Dimensional Space-Time

1.1 Determination of the Number of Space-Time Dimensions

The determination of the dimensions of a space-time is guesswork unless a principle(s) is used to specify the dimensions.

This chapter derives the dimensions of our 4-dimensional space-time based on two approaches that yield a dimension of four. The first approach uses a principle (axiom) that uses perhaps the most fundamental and general logic construct, qubits, to determine the dimension of space-time based on the axiom (assumption) that the number of fundamental interactions equals the dimension of space-time.[1] The second approach is based on the requirement that space-time must allow physical processes to run in parallel.[2] We view this requirement as an axiom.

Both approaches lead to a space-time dimension of four for our space-time and a spatial dimension of 192 for the Megaverse.[3].

1.2 The Determination of the Dimension from the Number of Interactions

We will assume that interactions have a dual role in fundamental physics: they determine the dynamics of particles, and they act to determine space-time dimensions.[4]

The motivation for the second role can be discerned from considering a 2-dimensional space, and introducing a simple 1/r potential such as:

$$V = g^2/(x^2 + y^2)^{-½}$$

where g is a coupling constant.

One can view V as the potential of a force. However the values of V suitably extended to the range [–∞, +∞] can be viewed as a third dimension. In addition, the basis of General Relativity is the role of gravity to determine the curvature of the universe. One can simply say that the universe is curved. One could also say that the universe is ‘curved’ into a surface in an implicit higher dimensional space-time.

These considerations lead us to propose the axiom:

[1] This approach was used in Blaha (2017c) and (2017d) to determine the dimensions of the Megaverse.
[2] This approach is developed in (2012a) and (2015a).
[3] Blaha (2017d).
[4] Blaha (2017c).

The number of interactions in a set of interactions equals the dimension of the space-time.

In section 1.3 we will determine the number of dimensions of our universe from the most fundamental construct in Logic and Reality.

1.3 The Determination of the Dimensions of Our Universe from the Only Essential Interactions

If we consider the universe as a thing in itself without reference to its matter content or their interactions there is only one construct that presents itself as an absolute necessity for a physical theory: Logic. In its most fundamental and most general form Logic is based on *qubits*. A qubit is a unit of quantum information that constitutes a state quantum system. Qubits are fundamental to Quantum Logic and, in a restricted form, furnish the truth values of true and false in 'classical' Logic.

Being a two-state system with the capability of having complex values and with values being able to be superposed, qubits can be viewed as having an associated U(2) unitary group whose operators serve to 'rotate' qubit values. Thus the U(2) group can be viewed as providing the 'interaction' that transforms qubits from one form to another.[5]

The U(2) qubit group has four generators. Taking the qubit as the fundamental construct of Reality since all Physics is based on Logic and ultimately Quantum Logic, we identify the source of our 4-dimensional space-time in the four U(2) generators. In chapter 3 we will show that the correct fundamental theory of Physics is a logical (mental) construct. With the choice made here we base Physics on Quantum Logic and the qubit interactions in particular.

In the next section we will show that this analysis is consistent with the seond approach based on the requirement (axiom) that physical processes must be able to execute in parallel. We thus achieve a unified framework for understanding the dimensionality of our space-time. From the dimension of space-time, and the minimal requirement that the dynamical evolution of physical systems needs a time coordinate, we are led to the Complex Lorentz group in flat space-time and Complex General Relativity in curved space-time.[6,7]

1.4 Asynchronous Logic and the Dimension of Space-Time

In this section we show that a consideration of the physical requirement (axiom) that Nature must allow physical processes to evolve dynamically in parallel leads to the principle of Asynchronous Logic that, in turn, necessitates a space-time dimendion of four.

Thus we have two approaches based in Logic that inexorably lead to 4-dimensional space-time.

---

[5] These operators have two conditions placed on them in quantum logic so they are analogous to electromagnetism with its restriction by gauge conditions to two degrees of freedom.

[6] With the further assumption that the speed of light is the same in all physical reference frames in flat space-time.

[7] We use similar arguments in the Megaverse (the universe of universes) discussion in Blaha (2017d) to determine the Megaverse dimension is 192.

1.4.1 Synchronization of Non-Local Physical Processes

In earlier books we discussed the central role of Logic and the need for synchronization of non-local physical processes. The need for synchronized non-local physical processes requires the introduction of a new principle: the Principle of Asynchronicity.[8] When processes take place in parallel whether it is Quantum Mechanical entangled processes at small/large distances, or in high order Feynman diagrams (or their old fashioned time ordered perturbation theory predecessor) the synchronicity of a process is a physical requirement. It is implicitly resolved by physical laws which prevent asynchronicities (situations when parallel processes get "out of sync" resulting in the failure of an entire physical process to complete properly.) The Principle of Asynchronicity is described in the following pages. Asynchronicity can be briefly described as:

> In computation asynchronicity issues can arise. For example parallel computations or computer processes on a chip or set of chips have to be carefully managed for a parallel computer process to complete properly. In the case of computer chip design (VLSI chips and so on) techniques have been developed for the design of chips based on multi-valued logic. One conceptual approach uses 4-valued logic to define clockless computer logic circuits. The 4-valued logic developed by Fant (2005) has the four logic values TRUE, FALSE, NULL, and INTERMEDIATE. It is an extension of Boolean Logic that can accommodate time asynchronicities in asynchronous computer circuits. It enables circuits to avoid the use of system clocks to implement synchronization.[9] Thus the synchronization is explicit in 4-valued logic and non-logical constructs are not needed.[10] Concurrent transitions are coordinated solely by logical relationships with no need for any time constraints or relationships.

Now, realizing that The Standard Model, and physical theories that are ultimately derived from it such as Quantum Mechanics, potentially contain asynchronicities, we suggest that a Principle of Asynchronicity is embodied in the fundamental theory of Physics that leads to Dirac-like equations for the fundamental fermions – the leptons and quarks of The Standard Model.

1.4.2 Four-Valued Asynchronous Logic

The basic defining features of asynchronous circuits and Asynchronous Logic are:

---

[8] Much of this chapter is covered in Blaha (2011c) and printed in smaller type. Some might argue that it should be called the principle of synchronicity since the goal is synchronization of the parts of an evolving process. We chose to follow the terminology in the field of Asynchronous Logic as exemplified by Fant (2005) – a classic in that field.

[9] Remarkably Bjorken (1965) pp 220-226 presents an analogy of Feynman diagrams with electrical circuits where momenta map to currents, coordinates to voltages, Feynman parameters to resistance, and free particle equations of motion to Ohm's Law plus the equivalent of Kirchhoff's Laws. Thus Feynman diagrams and computer circuits are completely analogous.

[10] A two-valued asynchronous logic is also possible – just as the Dirac equation can be expressed as two 2-dimensional equarions. See Fant (2005) and Bjorken (1965).

1. An *asynchronous circuit* is a circuit in which the component parts are autonomous and can act in parallel at various rates of time evolution. They are not controled by a clock mechanism but proceed or wait for signals indicating that they can proceed.
2. *Asynchronous logic* is the logic used in the design of asynchronous circuits. The logic embodies the asynchronicity, and so the circuits built using the logic do not use a clock to control the execution speed of the various parts of an asynchronous circuit. Consequently logic elements do not necessarily have a distinct true or false state at any given point in time. 2-valued Boolean Logic is not sufficient and so asynchronous logic is multi-valued. The logic embodies states that allow for "stop and go" states within an executing asynchronous circuit.

In Fant's asynchronous 4-valued logic the four possible truth values of a state are:

True – status is true and all data is current
False – status is false and all data is current
Intermediate – status is indefinite with some data current
NULL – status is indefinite with no data present – results in a suspension of processing of the circuit part in a NULL state until current data becomes present

"Data" is the information flowing through all or part of a circuit. Using these truth values the evolution in time of the parts of an asynchronous circuit are effectively synchronized by the logic without the use of a clock mechanism. (A clock mechanism effectively is a subsidiary time constraint or set of time constraints.) See Fant (2005) for further details.

An implicit aspect of asynchronous logic is the coordination of spatially separated parts of a circuit. Since spatial separations in a circuit can be mapped to time delays using the speed of data propagation between parts, spatial asynchronicites are subsumed under time asynchronicities. This is particularly true for computer chips which are kept small to minimize delays.

1.4.3 Principle of Asynchronicity

An obvious feature of elementary particle phenomena is the coordination of the parts of a physical process in time and space. Complex Feynman diagrams embody the coordination of the parts of interacting particles. Quantum entanglement phenomena embody the coordination of the parts of a physical phenomena separated by large distances and perhaps times. Examples of these types, which could be multiplied indefinitely, lead to a Principle of Asynchronicity.

Principle: Nature requires asynchronicity. This asynchronicity is coordinated by 4-valued physico-logical structures for matter.

Elaboration: Elementary particle physical phenomena must support extended coordinated physical phenomena in space and time. The fundamental laws of particle physics must be such as to permit coordinated physical phenomena with coordination between the parts of a physical phenomenon at small/large distances and small/large time intervals. The coordination must be embodied within physical laws.

This principle will be applied below to justify Dirac-like equations for particle dynamics.

Coordination is an obvious feature of physical phenomena. This principle goes beyond that by asserting that extended coordinated physical phenomena must exist. If particles exist, then their antiparticles must also exist to provide asynchronous behavior in interaction regions. If only particles existed then all interactions would proceed forward in time and the state of the interaction at any point in time would be known. With the addition of antiparticles, asynchronicity issues are introduced and at various 'time slices' (if one thinks in terms of old-fashioned time ordered perturbation theory) of the progress of an interaction, the state can be ambiguous since antiparticles are negative energy particles moving backward in time.

Asynchronicities are common in the many subcircuits of a computer chip. Asynchronicities are also common in the many interaction subregions of a set of particles in interaction. Page 7 of Fant (2005) has a diagram of a circuit with a set of subcircuits with five time slices of the interacting subcircuits showing five states of the "'data' wavefront" at five points in time. This diagram is similar to the time-sliced diagram of an interacting system of particles in "old fashioned" time-ordered perturbation theory. Page 29 of Blaha (2005b) displays a similar diagram (Fig. 5.1.4) in a description of a Standard Model Quantum Langauge Grammar – a language representation of particle physics. Blaha's diagram[11] is remarkably similar to Fant's diagram in overall features as one might expect since both address time asynchronicity.

The asynchronicity that appears in perturbation theory diagrams is intimately related to the appearance of antiparticles in diagrams. As noted earlier antiparticles are interpretable as negative energy particles traveling backwards in time. The time orderings, which are implicit in the Feynman diagram approach and explicit in old fashioned perturbation theory, evidence time asynchronicity and the effects of the dynamics. They coordinate asynchronicities such that correct results follow from perturbative calculations.

1.4.4 Why add Space to Logic?

Space is necessarily a part of physical theory because propositions often depend on a spatial location. Thus we must add spatial (and time) dimensions to our specification of the physical Reality as well as Theory. Later we will connect the spin ½ matrix formulation of Operator Logic[12] augmented with space-time dimensions to the Dirac equation for spin ½ particles. Thus we map Operator Logic spinors and the spinors in physical Reality. Fermion particles (spin ½ particles) in physical Reality emerge from a map from Operator Logic spinors.

Now we address the issue of the number of space-time dimensions. Clearly if spinors exist in Reality, as we know they do, then they must be "spinning" in spatial dimensions. The number of components of a spinor is related to the total number of time and space dimensions.[13] For the case of an even number of dimensions d a spinor has $2^{d/2}$ components. For the case of an

[11] Created without knowledge of Fant's work.
[12] See Blaha (2010a).
[13] Weinberg (1995) p. 216.

odd number of dimensions d a spinor has $2^{(d-1)/2}$ components. Based on these formulas we find the results in the following Table 1.1.

The case of d = 1 is immediately ruled out because Operator Logic supports, at minimum, 2 component spinors or 4 component spinors. The case d = 2 is also ruled out because spinor particles in a one-dimensional space reduce to scalar particles, and Reality has true spinors. The case d = 3 is ruled out because in two spatial dimensions there is no difference between left-handedness and right-handedness. *Thus the minimal number of spatial dimensions that yield true physical spinors and support "handedness" is three spatial dimensions.* This case meets Leibniz's Decision Axiom (chapter 3). The simplest features associated with space are spin (represented by spinors) and handedness. They yield a rich spectrum of particle types and interaction types (maximal complexity).

Total Number of Space-Time Dimensions d	**Number of Spinor Components**
1	1
2	2
3	2
4	**4**

Table 1.1. The number of spinor components for various numbers of space-time dimensions.

Thus we have a rationale for the extension of Operator Logic to include one time and three spatial dimensions.

The Principle of Asynchronous Logic and the parallel nature of Physical processes in general imply our space-time dimension is 4.

1.4.5 Truth Is Generally Local – Space-Time Dependent

The extension of Operator Logic to include space-time is further buttressed by the dependence of the truth of statements on location and time in general. For example: "Today it rained in Concord, New Hampshire." is a space and time dependent statement.

Thus we find that statements are *local* in general – they depend on the time and spatial location.

The locality that we find in Logic naturally leads to locality in physical theories – particularly the locality of the Yang-Mills rotations in quantum field theories such as The Standard Model where the values of quantum numbers can vary from space-time point to space-time point but in such a way that their variation is compensated by local rotations of quantum fields. The locality of logical statements thus supports the connection of Logic to fundamental Physics.

1.4.6 Matrix Representation of Asynchronous Logic

The four possible logic states of Asynchronous Logic can be mapped to a matrix representation with four component columns and 4×4 matrices that transform between logic values.[14] The basic four pure logic states can be labeled using a notation that connects with physics:

$$u(+\tfrac{1}{2}) = \begin{bmatrix} 1 \\ 0 \\ 0 \\ 0 \end{bmatrix} \qquad u(-\tfrac{1}{2}) = \begin{bmatrix} 0 \\ 1 \\ 0 \\ 0 \end{bmatrix}$$

and

$$v(-\tfrac{1}{2}) = \begin{bmatrix} 0 \\ 0 \\ 1 \\ 0 \end{bmatrix} \qquad v(+\tfrac{1}{2}) = \begin{bmatrix} 0 \\ 0 \\ 0 \\ 1 \end{bmatrix} \tag{9.2}$$

The arguments of u and v will become physically values of particle spin: +½ represents an "up" spin state and –½ represents a "down" spin state. Linear combinations of these four states obtained using linear combinations of the sixteen 4×4 matrices with complex coefficients form *qubits*.[15] Any bit or qubit transformation can be constructed from a sum of the sixteen Dirac matrices[16] multiplied by complex coefficients. The states listed above are constants. They will become variable through the use of Lorentz transformations considered later.

A set of sixteen independent 4×4 matrices can be constructed from the four basic Dirac matrices, usually denoted γ^{μ}, by multiplications and summations. They can be found in most books on quantum field theory.

Applying these basic facts about the 4×4 matrix representation of Asynchronous Logic to particles we can develop the Dirac equation formulation of spin ½ particles after introducing space-time coordinates.

We can view fermion particles as parametrized by coordinates and interacting via interactions (forces) that primarily result from the need to impose symmetry requirements on coordinate transformations.[17]

---

[14] See Blaha (2010a) for more details.

[15] S. Weisner, "Conjugate coding". Association for Computing Machinery, Special Interest Group in Algorithms and Computation Theory **15**, 78–88 (1983). We will not discuss qubits in this construction.

[16] See Bjorken (1964).

[17] We note that Asynchronous Logic can also be formulated with two dimensional vectors and 2×2 matrices just as the Dirac matrix formalism for spin ½ particles can be expressed in a 2×2 matrix formalism.

1.5 Matter is Insubstantial Neglecting Particle Interactions

Philosophers and physicists have debated the nature of matter for milleniums. Differing opinions have been the norm. Perhaps one of the most interesting expressions of opinion was that of Dr. Johnson, the 19th century author of the Dictionary. Upon hearing of Bishop Berkeley's philosophic view that matter was insubstantial and "not real", Dr. Johnson proceeded to kick a rock while exclaiming "I refute it thus" according to his biographer Boswell. While Dr. Johnson's riposte cannot be denied in its succinctness, our theoretical development of fermion particles, and the further development of our construction, later, leads to a refinement of the Berkeley-Johnson conflict as well as that of many philosophers and physicists.

For we propose that matter is truly insubstantial with Bishop Berkeley, and yet gains substantiality through interactions (forces) without which all matter would be interpenetrable and could reside at a single point.[18] Thus Dr. Johnson does not truly refute Bishop Berkeley but does show that forces exist amongst matter that gives it "substantiality."

Since all matter might have been concentrated[19] at an 'essentially' mathematical point in the absence of interactions we can call that point the *void* and view it as the 'location' of the Big Bang that presumably existed at the Beginning. Coordinates and interactions may well have originated at that point as the result of quantum fluctuations.

---

[18] To some extent we see an approximation of this proposal in the approximate interpenetrability of normal matter and Dark Matter which would be exact if there were not a very weak force between matter and Dark Matter as well as the gravitational interaction.

[19] This comment is based on our version of quantum field thory in which all interactions (including gravity) go to zero at zero distance. See Blaha (2005a) for a detailed discussion of our divergence-free form of quantum field theories. With zero forces, all particles can concentrate at a single point.

2. Unified SuperStandard Model and its Hidden SuperString Aspect

The Unified Standard Model that we have developed is a type of Quantum Field Theory that contains the known Standard Model and Gravitation as well as significant new features. It uses extensions of Quantum Field Theory: a Two-Tier PseudoQuantum formulation that makes the theory finite to all orders in perturbation theory and can be quantized in any coordinate system.

In this chapter we will show that the theory has a String aspect that, together with its SuperSymmetric aspect, enables us to characterize it as a hidden variant of SuperString theory.

2.1 String Theory vs. Two-Tier Quantum Field Theory

We begin by comparing the Gervais-Sakita[20] SuperString lagrangian (GS) with a simple Two-Tier Quantum Field Theory lagrangian:[21]

$$\mathcal{L} = \overline{\psi}(x)(i\gamma^{\mu}\partial/\partial x^{\mu} - m)\psi(x) - \partial_a X_{\mu}\partial^a X^{\mu} \qquad \text{(GS)}$$

$$\mathcal{L} = \overline{\psi}(X_T)(i\gamma^{\mu}\partial/\partial X_T{}^{\mu} - m)\psi(X_T) + \tfrac{1}{4} M_c{}^4 F^{\mu\nu}(X_T)F_{\mu\nu}(X_T) \qquad \text{(TT)}$$

where

$$X_{T\mu}(y) = y_{\mu} + i\, Y_{\mu}(y)/M_c{}^2$$

and

$$F_{\mu\nu} = \partial X_{\mu}/\partial y^{\nu} - \partial X_{\nu}/\partial y^{\mu}$$
$$\equiv i\,(\partial Y_{\mu}/\partial y^{\nu} - \partial Y_{\nu}/\partial y^{\mu})/M_c{}^2$$

The Gervais-Sakita lagrangian differs from the Two-Tier lagrangian by the introduction of the 'string' X^{μ} in the fermion part of the Two-Tier lagrangian. We note that the Two-Tier quantum field Y_{μ} has two degrees of freedom like the string field.

Thus the Two-Tier fermion field is a function of a string[22] whereas the SuperString lagrangian treats the fermion and the string as independent. As a result the Two-Tier quantum

---

[20] Gervais, J. L. and Sakita, B., Nucl. Phys. **B34**, 632 (1971).
[21] Blaha (2002) and (2005a).
[22] As Blaha (2002) and (2005a) show there is a form of the Calculus of Variations that supports a canonical derivation of the equations of motion.

field theory embedded in the Unified SuperStandard Model is a variant on SuperString theory if one also takes account of the SuperSymmetry attributes of the Unified SuperStandard Model.

2.2 Strings in Particles? or Strings Made into Particles?

A comparison of the two two types of theories is made more pointed by the form of Feynman propagators in Two-Tier theories. Fig. 4.1, which is discussed in detail in section 4.2.6 and displayed below because of its relevance, shows that a 'dressed' free fermion particle has an associated cloud of 'strings' that are the source of the finiteness of perturbation theory calculations. Thus Two-Tier Unified SuperStandard Model calculations are finite to all orders in perturbation theory by putting 'strings' in particles unlike SuperString theories which make particles from strings.

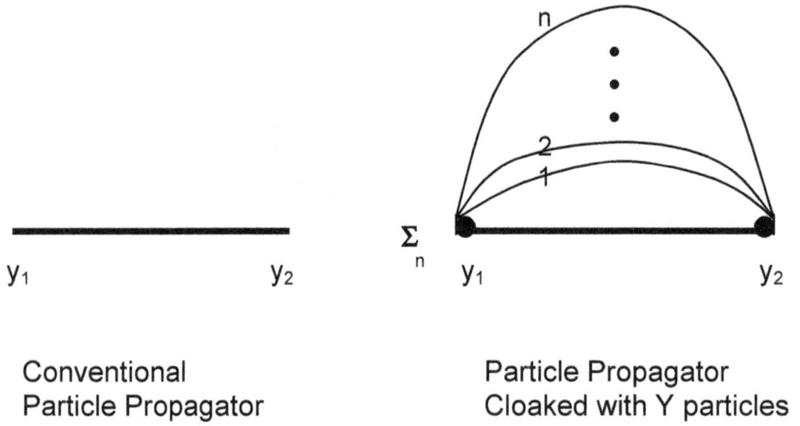

Figure 4.1. Feynman diagram for conventional and cloaked Two-Tier propagators.

2.3 Further Evidence of the String-like Substructure of the Unified SuperStandard Model

Quantum Dimensions $X_T^\mu(y)$ endow a particle with an extended structure that resembles to some extent the extended structure seen in bosonic string and Superstring theories. For example, Bailin (1994) use the operator[23]

$$V_\Lambda(k) = \int d^2\sigma \sqrt{-h}\, W_\Lambda(\tau, \sigma)\, e^{-ik\cdot X}$$

where X_T^μ is a quantized fourier expansion of the string fields (see eq. 7.22 of Bailin (1994)).

[23] D. Bailin and A. Love, *Supersymmetric Gauge Field Theory and String Theory* (Institute of Physics Publishing, Philadelphia, PA, 1994) page 272.

We note our X_T^{μ} coordinate-field has two transverse degrees of freedom due to gauge invariance, which also invites comparison to the bosonic string. A point of difference is that we will create a well-defined quantum field theoretic formulation in conventional space-time that has the Standard Model as its "large distance" behavior thus introducing a note of reality that is not (yet?) very apparent in Superstring theories. We see that the interacting quantum field theories based on this approach also have good, finite, short distance behavior just as string theories.

Scalar, and other particles', Feynman propagators can be viewed as describing the propagation of a particle cloaked (accompanied) by a cloud of Y particles

3.The Three Players in Creating a Fundamental Theory of Physical Reality

When one considers constructing a theory, or model, of fundamental 'real world' physical reality, the usual course of action is to make some assumptrions such as a symmetry group, or a construct like strings, and then proceed to develop their implications.[24] Next, the authors of the theory, and experimenters, study the experimental implications of their theory. A tacit assumption, which is sometimes stated, is that their mathematical-mental construct is somehow automatically implemented by Reality. The possibility that their construction is merely a mental exercise, and not Reality, is seldom considered except by the philosophically inclined. And for good reason. No one has put forward a concept that joins thought to Reality in a convincing way except as 'word play.'

In this chapter we will seek to carefully characterize the process of constructing a fundamental physical theory.[25] In doing so, we identify 'three players' that play decisive roles in bringing a fundamental theory from conception to Reality. First is the construction process which is a mathematical-mental effort that leads to a complete theory fundamental theory. Secondly, there is the conceptual mapping of the theory to physical Reality. Thirdly, there is the 'agency' by which Reality is imprinted with the theory. The third 'player', the 'imprinter,' is often alluded to in the statement that it is surprising how well Reality is described by mathematics. This statement is not good enough! We will examine it in some detail in this chapter together with the other two players in the 'game' of constructing the fundamental theory of everything.

3.1 Construction of a Fundamental Physics Theory of Everything

The construction process of a fundamental theory is a mental exercise consisting of defining primitive constructs, specifying assumptions (axioms), and then proceeding to develop a theory from the assumptions.

[24] In the author's view this approach is satisfactory for non-fundamental theories and models where the underlying physics is presumably well understood. But at the fundamental physics level a more careful analysis is required. This chapter delves into this issue in some detail.

[25] We discuss the metaphysical and logical basis of our approach in more detail in Blaha (2011c).

3.1.1 Euclidean Geometry Model

Euclid's theory of geometry is perhaps the best model of the process, and the components of the process. First Euclid identifies primitive unspecified constructs: straight lines, curves, and angles (which are viewed in no need of definition.) He then states five axioms and proceeds to state and prove theorems based on the unspecified, but understood, properties of the constructs. As the set of theorems grows, subsidiary, but unspecified, properties of geometric figures are used in proofs. These properties include the definition of angles, the construction of simple geometric figures such as rectangles, triangles, trapezoids, and so on. The more advanced developments in Euclidean geometry such as the geometry of Ptolomaic astronomy[26] with cycles and epicycles also have implicit principles for construction.

We see from Euclid that the theory construction process requires:

1. The definition of primitive terms.
2. The specification of a set of axioms.
3. The derivation of theorems from the axioms and further ad hoc but necessary assumptions.
4. The identification of the 'complete' theory with Reality.

Euclid was fortunate in the creation of Euclidean geometry because he knew the method with which to construct geometry, and its primitive terms with which to define his axioms, and then his theorems. The development of a fundamental physical theory is not so fortunate.

3.1.2 Logic Applied to Physics

Perhaps the most important aspect of constructing a physical theory is the use of Logic to derive results from the fundamental axioms. Questions have been raised about the validity of Logic. Attention is often drawn to logical 'paradoxes' and particularly to Gődel's Theorem. In Blaha (2015a) we showed the problems of these paradoxes and Gődel's Theorem were due to the choice of subjects in a statement that were not in the subject domain of the statement. Statements are a verbal form of functions. Like functions they have a domain (set of valid subjects) and a range.

Most scientists, Logicians and Mathematicians are familiar with mathematical functions that have numerical arguments and calculate a numerical value or set of numerical values. However, those familiar with Computer Science know that a function, in general, has two types of output: its numeric value(s), and a status value indicating whether the computation that the function performed was successfully done or failed. Typically the status value is a zero or one,

[26] Ptolomaic Astronomy is a very-well developed, and mathematically sound, branch of Euclidean geometry. Its flaw is its misidentification with the physical reality of the Solar System. In this regard it very well illustrates our point that the matching of theory with Reality is an important part of the creation of a fundamental theory.

but it is understood as true or false also. (True – computation successful; false – computation failed for some reason) Thus we categorize functions as being in one of three broad categories:

Types of Functions

1	2	3
A Mathematical Function Producing A Value(s) Only;	A Function Producing a Value(s) and a Status Value;	A Function Producing a Status Value Only (T or F);
Examples:		
sin function	Programming Function	Logic Statement

Thus Logic is unsullied by paradoxes and, besides being the only way to 'do' Physics theory, is correct and consistent.

3.1.1 Primitive Terms of Fundamental Physics Theories

The primitive terms of the fundamental theory of Physics is an open question. There are various theories that have been proposed: Twistor theory, SuperString theories, SuperStandard Model theory, and so on. They differ significantly in their definition of primitive terms and their axioms (or their equivalent to axioms).

Under the circumstances probably the most significant aspects of terms are:

1. Consistency with what we know of Reality – they are somehow evident in Reality.
2. Simplicity.
3. Utility in defining meaningful axioms. This feature implies a forward-looking approach that looks to the development of the theory to guide the choice of primitives.

Euclid's specification of primitive terms agrees with these considerations.

3.1.2 Physical Axioms of a Fundamental Theory

The choice of axioms is very dependent on the eventual theory. Generally it seems that most theories choose a lagrangian and/or symmetry group, and proceed to 'back-engineer' a set of axioms if they consider defining axioms at all.

If one wishes to create a fundamental theory using a top-down approach then the set of axioms, like primitive terms, should satisfy the same three conditions plus Decision Axioms (described below):

1. Consistency with what we know of Reality – they are somehow evident in Reality.

2. Simplicity.
3. Utility in generating the fundamental theory. This feature also implies a forward-looking approach that looks to the development of the theory to guide the choice of axioms.
4. The choice of axioms should be in agreement with the below Decision axioms.

3.1.3 Decision Axioms for a Fundamental Theory

The concept of Decision Axioms does not seem to have been formally considered previously. Decision axioms guide a choice between alternatives in the construction of fundamental theory. Given the complex nature of fundamental theories choices between alternatives frequently occur. SuperString theory, for example, has millions of choices in the determination of the 'correct' SuperString theory.

What rules (Decision Axioms) can guide a choice?

In this subsection we identify six rules that provide guidance for making the proper choice.

3.1.3.1 Ockham's Razor

William of Ockham proposed a Law of Parsimony that is called *Ockham's Razor* which states that the simplest choice is to be preferred in a multiple choice situation. This principle is often stated as 'the simplest solution is usually the correct solution.'[27]

The best rationale for this principle is that it generally reduces the complexity that follows such a choice. Since many physics calculations are extremely difficult, picking the simplest choice.would generally tend to make subsequent theorems/and calculations less difficult. This point of view might be thought to be ad hoc or anthropomorphic. But it reflects the reality of scientific calculation and of theory construction.

Thus we will assume Ockham's Law of Parsimony in the construction of our theory with the proviso that Leibniz's Minimax Principle (below) takes precedence if there is a conflict in the implications of the choices.

3.1.3.2 Leibniz's Minimax Principle for Physics Theories

Leibniz[28] developed a Minimax Principle that can be phrased for our purposes as, "The universe is based on the smallest set of properties or features that lead to the greatest variety of phenomena." This principle reflects the spirit of the minimum/maximum criteria of the Calculus of Variations[29] that plays a central role in many physics theories. This principle somewhat

---

[27] William of Ockham – Law of Parsimony – "Pluralitas non est ponenda sine necessitate" or "Plurality should not be posited without necessity." Ockham's Law was first stated by Durand De Saint-Pourçain (1270-1334 A.D.). In simple terms the principle states the simplest solution to a problem is most likely to be the correct solution.

[28] See Rescher (1967).

[29] Leibniz was one of the founders of the Calculus of Variations.

overlaps Ockham's Law of Parsimony. Given a choice of possible theoretical lines of construction there is a possibility that the Law of Parsimony and the Minimax Principle would suggest different choices. Fortunately, we will see that the construction of the SuperStandard Model does not seem to present this potential dilemma.

An important, unremarked, aspect of Leibniz's Principle is the decision between a set of choices depends on the future part of the construction or theory. Thus future constructs determine past constructs in minimax decisions.[30]

3.1.3.3 Conputationally Minimal

Given a choice between two alternatives that appear equally likely from general considerations and the view of the preceding two Decision Axioms, the preferred choice is the one leading to the most minimal (simplest) computations in the full theory.

3.1.3.4 Experimentally Verifiable Directly or Indirectly

Given a choice between two alternatives, one of which is subject to experimental verification, directly or indirectly although perhaps in the future, and the other choice is not, then the first alternative is preferred.

3.1.3.5 Level of Rigor

Physics is mathematical in nature. Mathematics strives for rigor and will not be satisfied without rigorous results except for conjectures and speculations. Physics prefers rigorously derived results as wel. However, there is 'physical rigor,' which is a level of rigor that is rigorous to the extent that it is mathematically rigorous. This circuitous statement reflects the historical fact that Physics has often gone beyond the mathematics of the time. The most significant example is Newtonian physics, which until the mid-19th century used *derivatives* asserting that they were, or would become, rigorous mathematically. After a careful analysis of derivatives, and advances in the understanding of continuity and the various types of infinities, mathematicians were able to put derivatives on a rigorous footing justifying the use of derivatives by physicists for almost four centuries.

Today, the most interesting part of physics in search of mathematical rigor is the path integral formulation in quantum theory and the Faddeev-Popov Method, in particular. The problem here is again an understanding of infinities in functional derivatives and path integrals. Again, mathematical rigor is lackng. Yet physicists use these techniques with the strong belief that their results will be eventually justified rigorously.

---

30 The knowledgable reader will remember Feynman's speculation that the physical universe may be evolving from the future into the past. Quantum field theories support such an interpretation of their mathematics. The similarity of this fundamental minimax principle's feature with the corresponding feature of physics theories encourages support for the minimax principle as a 'design' law of fundamental physical theory.

In view of the nature of rigor in Physics it seems reasonable to state the Decision Axiom: *In the case of a situation where several possible approaches are possible, one should choose the approach that is most rigorous – should one exist. If all approaches are at an 'equal'* level of rigor then the approach that appears 'most' physical should be chosen, taking account of the other Decision Axioms.

3.1.3.6 Nature Tends to Repeat Successful Strategies

In many branches of Science one sees that successful strategies and techniques are repeated in several areas – perhaps with some variations and changes. Consequently, when a physical phenomena is similar to another physical phenomena, and all other axioms that were stated above are not relevant for the phenomena, then one should model the new phenomena in a manner similar to the successfully modeled phenomena.

3.1.4 The Body of the Derived Theory –Jeopardy Points

After the primitives and the axioms have been specified, the body of the theory can be developed mathematically (logically). At points in the development process decisions may have to be made between alternatives. We call these points – *jeopardy points*. The above Decision Axioms can provide guidance if a strictly physical decision is not evident.

Clearly the fewer the jeopardy points in a theoretical development, the better. The existence of many such points raises the question of whether the axiomatic basis needs to be strengthened.

The general lack of jeopardy points suggests the axiomatic basis of the theory is well-founded.

3.1.5 The Fundamental Theory as a Mental Construct

If one creates a fundamental theory it is a mental construct, and although physicists are quick to identify it with Reality, it is not Reality until a map to Reality is specified. Parmenides (front cover) pointed out this distinction in a way in his only surviving poem *On Nature*. He said the fundamental nature of the universe was an unchanging unity of nature (The Way of Truth); but our view of the universe was that of change and seeming (The Way of Appearance). Translated into the framework of our discussion we see the fundamental theory as an unchanging mental construct, but the Reality to which it maps as ever changing.

In the next section we discuss the mapping realizing that most physical theories make the leap from concept to Reality without effort.

3.2 Conceptual Mapping of a Fundamental Theory to Physical Reality

Generally the fundamental theories of Physics that have been proposed are so close to Reality that the mapping process is straight-forward – pick a symmetry group, specify a lagrangian, find the physical particle spectrum (which was put there by the choice of symmetry group anyway), and calculate experimentally verifiable numbers.

However in some cases the mapping is not so clear. For example, one might find an extremely large particle spectrum with experimentally measurable masses, and the particles are not there! Then one has the dilemma of redesigning the theory.

3.2.1 Mapping of the Fundamental Physical Axioms

One could possibly redesign the set of axioms of the theory if they were at fault and if the axioms had been specified. The new theory that emerges might then be mappable to experimental reality.

3.2.2 Mapping of the Body of the Derived Fundamental Theory

If the body of the fundamental theory does not agree with a mapping to Reality, the creator(s) of the theory often change the parameters or otherwise tinker with its development. One sees this in some current theories.

3.3 The 'Agency' by which Reality is Imprinted with a Fundamental Theory

Mental processes such as the thought embodying a physical theory are clearly different from physical reality. A question that has been a subject of discussion for over 2500 years is How does thought become embodied in Reality? A further question that has been of great interest for the past five hundred years is How does the mathematics of physical theory so well explain/correspond to Reality?

Addressing the second question we note that there are two possibilities: Reality is chaotic or Reality follows physical laws (which are necessarily mathematical). If Reality were chaotic but through some principle of self-organization transformed itself into an ordered system, then we would see order but there would be pockets of chaos in the universe and on earth that we have not seen. Further the transition to order would have to repeat with every change of state of a physical system – an effect which is also not seen. Thus 'self-organization' in any form does not appear to be the source of Physical Law.

Thus we must conclude that nature is ordered and conforms to physical laws. Physical laws are intertwined through mathematics. The fact that we can define theories that embody all physical laws – rather than have a host of unrelated physical laws is another marvel.

So we are left with the issue of who imprinted the concepts of a fundamental physical theory on Reality. The philosophers (Presocratic and more recent) addressed this problem and came to the conclusion that there was an 'Unmoved Mover' – an entity that implemented the imprinting of physical law on Reality.

Most philosophers (and theologians) identified this entity as God. And they called this the proof of the existence of God.

4. Unified SuperStandard Model Derivation

Chapter 3 outlined the process of deriving a unified theory of everything from a set of axioms. In this chapter we will outline that process for our Unified SuperStandard Model.[31] Our theory has the somewhat unique feature of being derived from first principles. Most other fundamental theories are constructed in an ad hoc fashion, often based on symmetry considerations, with features being added rather than derived.

The goal of this chapter is to derive the SuperStandard Model in the manner of Euclid with a clear connection between the steps of the derivation just as Euclid developed geometry from a progression of theorems.

4.1 Primitive Terms and Axioms

Primitive terms can be as simple as those of Euclid or they can be more complex. The level of simplicity depends on the nature of the theory and the Physical Laws that emerge from it. In the case at hand, a fundamental unified theory, the constructs that emerge in the construction of the theory are mathematically complex. Consequently, the choice of primitive terms and axioms must be expected to be mathematically complex as well unless one wishes to expand the primitive terms into an extensive term by term description in simpler, more basic primitives. We will not pursue that alternative here since the terms that we use are 'self-explanatory' to the Elementary Particle Physics theorist knowledgable about quantum field theory and particle symmetries.

4.1.1 Primitive Terms for the Unified SuperStandard Model

The primitive terms of the theory are:

Bits
Speed of Light
Space and Time Dimensions
Space and Time Coordinates
Covariance
Asynchronous
Parallel Processes

---

[31] It is described in detail in Blaha (2017c) and (2017d) as well as in versions in earlier books.

Reference Frame
Complex Lorentz Group
General Coordinate Transformations
Particle Masses
Fermions
Bosons
Particle State
Particle Rest State
Particle Momenta
Spin
Parity
Canonical Quantization
Quantum Field Theory
Asymptotic Particle State
Internal Symmetries
Coupling Constant
Discrete Symmetries
Yang-Mills Local Gauge Theory
Gravity
Universe
Surface Tension
Megaverse

In choosing these primitives, we understand that they each embody a significant theoretic description or body of knowledge. However their meaning is clear to the experienced theorist. We do not include names used in the mapping to reality (such as quark) in the list of primitives since the mapping to reality is a separate issue in our view.

4.1.2 Axioms for the Unified SuperStandard Model

The axioms that we will discuss below include the Decision Axioms of section 3.1.3. The physical axioms are

1. The dimensions of space-time are determine by the number of interactions.
2. Asynchronicity Principle supporting parallel processes is required.
3. Constancy of the speed of light in all reference frames.
4. Covariance for flat space-times of symmetry under the Complex Lorentz group of transformations.
5. Covariance for curved space-times of symmetry under Complex General Coordinate transformations.

6. Physically acceptable reference frames have real-valued coordinates. These coordinates can be obtained by transformations from complex-value coordinate systems.
7. Fermion and Boson vacua can be defined that are valid in all coordinate systems.
8. The number of particles in an asymptotic state of any given type is invariant in all reference frames.
9. Free fundamental fermions must have a real-valued energy.
10. All fields must be canonically quantized.
11. All interactions have a local Yang-Mills gauge theory formulation.
12. Gravity can cause space-time to be curved.

4.2 The Derivation of the Unified SuperStandard Model

The derivation of the Unified SuperStandard Model has been a multi-year process undertaken by the author. Much of the derivation appears in Blaha (2015a), (2016f), (2017b), (2017c) and (2017d). Earlier work is referenced in these books and listed in the References in this book.

In this section we outline the derivations presented in detail in these earlier books with some changes. The goal is to show a clear logical development of the Unified SuperStandard Model from first principles in a manner reminiscent of the derivation of Euclidean geometry. This derivation will be seen to be based on a 'simple' concept – the origin of space-time in our universe. The derivation at least explains the construction process of the physical theory. The manner of the derivation embodies the mapping of the theory to physical reality. The question of the 'Unmoved Mover' is necessarily beyond the scope of Physics.

4.2.1 Space-Time Dimensions

The number of space-time dimensions (4) is determined by the number of interactions. In the case of our universe it is set by the U(2) rotation group of qubits. See chapter 1 for details.

In the case of the Megaverse it is set to 192 by the number of internal symmetry interactions in the Unified SuperStandard Model.

4.2.2 Four Fermion Species

Under transformations of the Complex Lorentz group[32] from a state of rest, a spin ½ fermion can have one of four forms which we call *fermion species*.[33] These forms (species) are Dirac fermion and tachyon fermions with real-valued momenta; and complexon (Dirac-like) fermions and tachyon fermions with real energies and complex-valued spatial momenta (subject to the condition that the spin, real part of the spatial momentum, and the imaginary part of the spatial momentum are orthogonal to each other.) The energy of each boosted fundamental particle must be real-valued in the free field case since free fundamental particles do not decay.

The number of distinct species is not changed by the extension of the set of coordinate transformations to Complex General Coordinate Transformations.

We map the four fermion species to charged leptons, neutrinos, up-type quarks, and down-type quarks respectively.

Comment: This derivation of four types (species) of fermions *directly* from Complex Lorentz group boosts from a rest state to a state with real-valued energy (thus giving stable fundamental fermions in the absence of interactions) is the only derivation of four fermion species that does not make any assumptions about internal symmetries.

---

[32] Complex Lorentz group boosts are require to obtain four species. It is one of our axioms and a requirement of proofs in axiomatic quantum field theory. See Streater (2000).

[33] Most of the material in section 4.2 was presented in -Blaha (2012a), (2015a) and (2017b) as well as earlier books.

4.2.3 Fermion Lagrangian Terms

We next construct lagrangian terms for the fermions in the absence of interactions (free fermions) from the boosted rest particle fields.[34]

4.2.3.1 Dirac Particles – Charged Leptons

The free Dirac equation and lagrangian describes free charged leptons.

4.2.3.2 Tachyonic Particles - Neutrinos

The Dirac-like equation for tachyons, which correspond to neutrino species fermions, is

$$(\gamma^{\mu}\partial/\partial x^{\mu} - m)\psi_T(x) = 0$$

The "missing" factor of i in the first term requires the lagrangian to be different from the conventional Dirac lagrangian in order for the lagrangian to be real. The simplest, physically acceptable, free spin ½ tachyon lagrangian density is:[35]

$$\mathcal{L}_T = \psi_T{}^S(\gamma^{\mu}\partial/\partial x^{\mu} - m)\psi_T(x)$$

where

$$\psi_T{}^S = \psi_T{}^{\dagger}\, i\gamma^0\gamma^5$$

The corresponding action is

$$I = \int d^4x \mathcal{L}_T$$

Appendix 3-B of Blaha (2007b) proves I is real. The Hamiltonian density is

$$\mathcal{H} = \pi_T\dot{\psi}_T - \mathcal{L} = i\psi_T{}^{\dagger}\gamma^5(\boldsymbol{\alpha}\cdot\nabla + \beta m)\psi_T = -i\psi_T{}^{\dagger}\gamma^5\dot{\psi}_T$$

Tachyon canonical quantization follows the normal procedure with one change. (See Blaha (2017b) for details.)

We can't complete the canonical quantization procedure of tachyons in the conventional manner by fourier expanding the quantum field and specifying anti-commutation relations for the fourier component amplitudes: the incompleteness of the set of plane waves, which are limited by the restriction $|p| \geq m$, causes the anti-commutator of the fields not to yield

---

[34] Chapter 3 of Blaha (2017b) and earlier books.

[35] The tachyon theory we develop follows directly from Lorentz boosts. Previous tachyon theories such as those of Feinberg and Sudarshan were not derived in this manner and are different from our tachyon theory.

a $\delta^3(x - x')$. Thus the strictly conventional approach fails to yield the required anti-commutation relations.

Other possible approaches such as: 1) decomposing the tachyon field into left-handed and right-handed parts and then second quantizing each part; and 2) second quantizing in light-front coordinates ($x^{\pm} = (x^0 \pm x^3)/\sqrt{2}$); also both fail.[36]

The only approach that does succeed[37] is to decompose the tachyon field into left-handed and right-handed parts and then second quantize in light-front coordinates. We successfully followed that procedure in prior books.

4.2.3.3 Complexon Fermions – Free Up-Type Quarks

The complexon fermion equation (for free up-type quarks) is

$$[i\gamma^0\partial/\partial t + i\gamma\cdot(\nabla_r + i\nabla_i) - m]\psi_C(t, \mathbf{x}_r, \mathbf{x}_i) = 0$$

where the fields are functions of complex coordinates x_c:

$$x_c = (t, \mathbf{x}_r - i\mathbf{x}_i)$$

where the grad operators ∇_r and ∇_i are with respect to $\mathbf{x}_r$ and $\mathbf{x}_i$ respectively, and where there is a subsidiary condition on the wave function

$$\nabla_r\cdot\nabla_i\, \psi_C(t, \mathbf{x}_r, \mathbf{x}_i) = 0$$

We will call particles satisfying the above equations *complexons*. These fermions are not tachyonic. They satisfy

$$\gamma\cdot\nabla_r\gamma\cdot\nabla_i\psi_C(t, \mathbf{x}_r, \mathbf{x}_i) = \gamma\cdot\nabla_i\gamma\cdot\nabla_r\psi_C(t, \mathbf{x}_r, \mathbf{x}_i) = 0$$

We note that the above equations are covariant under the real Lorentz group and can be easily put into covariant form since the difference of the 4-vectors squared is a real Lorentz group invariant: $[\gamma^0\partial/\partial t + \gamma\cdot(\nabla_r + i\nabla_i)]^2 - [\gamma^0\partial/\partial t + i\gamma\cdot(\nabla_r - i\nabla_i)]^2 = 4\nabla_r\cdot\nabla_i$. *The complete details can be seen in Blaha (2017b) and earlier books.*

4.2.3.4 Tachyonic Complexons – Free Down-Type Quarks

Free tachyonic complexons (faster-than-light complexons – down-type quarks) satisfy

$$[\gamma^0\partial/\partial t + \gamma\cdot(\nabla_r + i\nabla_i) - m]\psi_C(t, \mathbf{x}_r, \mathbf{x}_i) = 0$$

---

[36] See the first edition Blaha (2006) where these possibilities were considered and found to fail.

[37] Blaha (2006) discusses this case in detail.

or

$$[\gamma \cdot \nabla - m]\psi_C(t, \mathbf{x_r}, \mathbf{x_i}) = 0$$

with the subsidiary condition

$$\nabla_r \cdot \nabla_i \, \psi_{CL}(t, \mathbf{x_r}, \mathbf{x_i}) = 0$$

We note that they are covariant under the real Lorentz group and can be easily put into (real Lorentz group) covariant form. *Again the complete details can be seen in Blaha (2017b) and earlier books.*

4.2.3.5 Comments

Again we note that the derivation of free Dirac-like dynamic equations and lagrangians follows *directly* from the Complex Lorentz group and the assumption that the action is real-valued. Thus we have the scaffolding for the introduction of interactions and the development of a complete SuperStandard Model.

4.2.4 Parity Violation

Parity violation has been a source of wonderment for over sixty years. We find the source of parity violation in the tachyonic neutrinos and down-type quarks in our theory. It follows directly[38] from the need to canonically quantize by first decomposing the tachyon field into left-handed and right-handed parts and then second quantizing in light-front coordinates.

The lagrangian terms that result, violate parity through the terms with tachyonic neutrino field factors and tachyonic down-type quark field factors. Thus the requirement of canonical quantization on our theory forces parity violation. In the following we use some equations from Blaha (2017b) (and our earlier books) that illustrate the origin of Parity Violation. Again, we see this feature follows directly from those of the previous subsection.

We can calculate the commutation relations of the left-handed and right-handed tachyon fields by pre-multiplying and post-multiplying by $\frac{1}{2}(1 - \gamma^5)$ and $\frac{1}{2}(1 + \gamma^5)$. The results are:

$$\{\psi_{TLa}^{\dagger}(x), \psi_{TLb}(x')\} = \tfrac{1}{2}(1 - \gamma^5)_{ab}\, \delta^3(x - x')$$
$$\{\psi_{TRa}^{\dagger}(x), \psi_{TRb}(x')\} = -\tfrac{1}{2}(1 + \gamma^5)_{ab}\, \delta^3(x - x')$$
$$\{\psi_{TLa}^{\dagger}(x), \psi_{TRb}(x')\} = \{\psi_{TR\alpha}^{\dagger}(x), \psi_{TLb}(x')\} = 0$$

where the subscript 'T' denotes tachyon field, 'L' denotes left-handed, and 'R' denotes right-handed. The † denotes hermitean conjugate, and 'a' and 'b' are spinor indices.

The free tachyonic lagrangian density decomposes into left-handed and right-handed parts:

$$\mathcal{L}_T = \psi_{TL}^{\dagger}\gamma^0 i\gamma^{\mu}\partial_{\mu}\psi_{TL} - \psi_{TR}^{\dagger}\gamma^0 i\gamma^{\mu}\partial_{\mu}\psi_{TR} - im[\psi_{TR}^{\dagger}\gamma^0\psi_{TL} - \psi_{TL}^{\dagger}\gamma^0\psi_{TR}]$$

Light-front variables can be defined by:

$$x^{\pm} = (x^0 \pm x^3)/\surd 2$$
$$\partial/\partial x^{\pm} \equiv \partial^{\mp} \equiv (\partial/\partial x^0 \pm \partial/\partial x^3)/\surd 2$$

with the "transverse" coordinate variables, x^1 and x^2, unchanged.

We define "+" and "–" light front, tachyon fields:

Left-handed, ± light-front fields: $\psi_{TL}^{\pm} = R^{\pm}C^{-}\psi_T$

[38] Blaha (2006) (and earlier books) discusses this subject in detail.

Right-handed, ± light-front fields: $\psi_{TR}{}^{\pm} = R^{\pm} C^{+} \psi_{T}$

Now if we transform to light-front variables and fields as above we obtain the light-front free tachyon lagrangian:

$$
\begin{aligned}
\mathcal{L}_T = {} & 2^{1/2}\psi_{TL}{}^{+\dagger} i\partial^{-}\psi_{TL}{}^{+} + 2^{1/2}\psi_{TL}{}^{-\dagger} i\partial^{+}\psi_{TL}{}^{-} - \psi_{TL}{}^{+\dagger}\gamma^0 i\gamma^j\partial^j\psi_{TL}{}^{-} - \psi_{TL}{}^{-\dagger}\gamma^0 i\gamma^j\partial^j\psi_{TL}{}^{+} - \\
& - 2^{1/2}\psi_{TR}{}^{+\dagger} i\partial^{-}\psi_{TR}{}^{+} - 2^{1/2}\psi_{TR}{}^{-\dagger} i\partial^{+}\psi_{TR}{}^{-} + \psi_{TR}{}^{+\dagger}\gamma^0 i\gamma^j\partial^j\psi_{TR}{}^{-} + \psi_{TR}{}^{-\dagger}\gamma^0 i\gamma^j\partial^j\psi_{TR}{}^{+} - \\
& - im[\psi_{TR}{}^{+\dagger}\gamma^0\psi_{TL}{}^{-} - \psi_{TL}{}^{+\dagger}\gamma^0\psi_{TR}{}^{-} + \psi_{TR}{}^{-\dagger}\gamma^0\psi_{TL}{}^{+} - \psi_{TL}{}^{-\dagger}\gamma^0\psi_{TR}{}^{+}]
\end{aligned}
$$

with an implied sum over j = 1, 2. In contrast to the light-front tachyon lagrangian above we note the corresponding light-front *"normal" Dirac fermion* lagrangian is

$$
\begin{aligned}
\mathcal{L}_{Dirac} = {} & 2^{1/2}\psi_{L}{}^{+\dagger} i\partial^{-}\psi_{L}{}^{+} + 2^{1/2}\psi_{L}{}^{-\dagger} i\partial^{+}\psi_{L}{}^{-} - \psi_{L}{}^{+\dagger}\gamma^0 i\gamma^j\partial^j\psi_{L}{}^{-} - \psi_{L}{}^{-\dagger}\gamma^0 i\gamma^j\partial^j\psi_{L}{}^{+} - \\
& - 2^{1/2}\psi_{R}{}^{+\dagger} i\partial^{-}\psi_{R}{}^{+} + 2^{1/2}\psi_{R}{}^{-\dagger} i\partial^{+}\psi_{R}{}^{-} - \psi_{R}{}^{+\dagger}\gamma^0 i\gamma^j\partial^j\psi_{R}{}^{-} - \psi_{R}{}^{-\dagger}\gamma^0 i\gamma^j\partial^j\psi_{R}{}^{+} - \\
& - im[\psi_{R}{}^{+\dagger}\gamma^0\psi_{L}{}^{-} + \psi_{L}{}^{+\dagger}\gamma^0\psi_{R}{}^{-} + \psi_{R}{}^{-\dagger}\gamma^0\psi_{L}{}^{+} + \psi_{L}{}^{-\dagger}\gamma^0\psi_{R}{}^{+}]
\end{aligned}
$$

The difference in signs between these lagrangians will turn out to be a crucial factor in the derivation of features of the Standard Model later.

Returning to the tachyon lagrangian we obtain equations of motion through the standard variational techniques:

$$
\begin{aligned}
2^{1/2} i\partial^{-}\psi_{TL}{}^{+} - \gamma^0 i\gamma^j\partial^j\psi_{TL}{}^{-} + im\gamma^0\psi_{TR}{}^{-} &= 0 \\
2^{1/2} i\partial^{-}\psi_{TR}{}^{+} - \gamma^0 i\gamma^j\partial^j\psi_{TR}{}^{-} + im\gamma^0\psi_{TL}{}^{-} &= 0 \\
2^{1/2} i\partial^{+}\psi_{TL}{}^{-} - \gamma^0 i\gamma^j\partial^j\psi_{TL}{}^{+} + im\gamma^0\psi_{TR}{}^{+} &= 0 \\
2^{1/2} i\partial^{+}\psi_{TR}{}^{-} - \gamma^0 i\gamma^j\partial^j\psi_{TR}{}^{+} + im\gamma^0\psi_{TL}{}^{+} &= 0
\end{aligned}
$$

This shows that $\psi_{TL}{}^{-}$ and $\psi_{TR}{}^{-}$ are dependent fields that are functions of $\psi_{TL}{}^{+}$ and $\psi_{TR}{}^{+}$ on the light-front where x^{+} equals a constant. They can be expressed in an integral form as well. (The independent fields $\psi_{TL}{}^{+}$ and $\psi_{TR}{}^{+}$ play a fundamental role in tachyon theory and are used to define "in" and "out" tachyon states in perturbation theory.)

The conjugate momenta are

$$
\begin{aligned}
\pi_{TL}{}^{+} &= \partial\mathcal{L}/\partial(\partial^{-}\psi_{TL}{}^{+}) = 2^{1/2} i\psi_{TL}{}^{+\dagger} \\
\pi_{TL}{}^{-} &= \partial\mathcal{L}/\partial(\partial^{-}\psi_{TL}{}^{-}) = 0 \\
\pi_{TR}{}^{+} &= \partial\mathcal{L}/\partial(\partial^{-}\psi_{TR}{}^{+}) = -2^{1/2} i\psi_{TR}{}^{+\dagger} \\
\pi_{TR}{}^{-} &= \partial\mathcal{L}/\partial(\partial^{-}\psi_{TR}{}^{-}) = 0
\end{aligned}
$$

The resulting canonical equal-light-front ($x^+ = y^+$) anti-commutation relations of the independent fields are:

$$\{\psi_{TL\,a}^{+\,\dagger}(x), \psi_{TL\,b}^{+}(y)\} = 2^{-1}[C^- R^+]_{ab}\,\delta(x^- - y^-)\delta^2(x - y)$$
$$\{\psi_{TR\,a}^{+\,\dagger}(x), \psi_{TR\,b}^{+}(y)\} = -2^{-1}[C^+ R^+]_{ab}\,\delta(x^- - y^-)\delta^2(x - y)$$
$$\{\psi_{TL\,a}^{+\,\dagger}(x), \psi_{TR\,b}^{+}(y)\} = \{\psi_{TR\,a}^{+\,\dagger}(x), \psi_{TL\,b}^{+}(y)\} = 0$$
$$\{\psi_{TL\,a}^{+}(x), \psi_{TR\,b}^{+}(y)\} = \{\psi_{TR\,a}^{+\,\dagger}(x), \psi_{TL\,b}^{+\,\dagger}(y)\} = 0$$

If we compare the above with the corresponding anti-commutation relations of *conventional* *<u>Dirac</u>* quantum fields:

$$\{\psi_{L\,a}^{+\,\dagger}(x), \psi_{L\,b}^{+}(y)\} = 2^{-1}[C^- R^+]_{ab}\,\delta(x^- - y^-)\delta^2(x - y)$$
$$\{\psi_{R\,a}^{+\,\dagger}(x), \psi_{R\,b}^{+}(y)\} = 2^{-1}[C^+ R^+]_{ab}\,\delta(x^- - y^-)\delta^2(x - y)$$

we see that the right-handed tachyon anti-commutation relation has a minus sign relative to the corresponding right-handed conventional anti-commutation relation. The right-handed tachyon anti-commutation relation with its minus sign will require compensating minus signs in its creation and annihilation Fourier component operators' anti-commutation relations.

The sign differences between the lagrangian terms above ultimately lead to parity violating features in the SuperStandard Model lagrangian and thus resolve the long-standing question:

Why parity violation? Answer: Nature preferentially chooses the Left-handed part of the complex Lorentz group..

The light-front, left-handed and right-handed tachyon lagrangian $\mathcal{L}_T$ and its equations of motion imply

$$H = i2^{-\frac{1}{2}}\int dx^- d^2x\,[\psi_{TL}^{+\dagger}\partial^-\psi_{TL}^{+} - \partial^-\psi_{TL}^{+\dagger}\psi_{TL}^{+} + \psi_{TL}^{-\dagger}\partial^+\psi_{TL}^{-} - \partial^+\psi_{TL}^{-\dagger}\psi_{TL}^{-} -$$
$$- \psi_{TR}^{+\dagger}\partial^-\psi_{TR}^{+} + \partial^-\psi_{TR}^{+\dagger}\psi_{TR}^{+} - \psi_{TR}^{-\dagger}\partial^+\psi_{TR}^{-} + \partial^+\psi_{TR}^{-\dagger}\psi_{TR}^{-} + \text{mass terms}]$$

After substituting for the various fields we find the *independent fields* (which create the in and out particle states) have the hamiltonian terms:

$$H = \sum_{\pm s}\int d^2p\,dp^+\,p^-[b_{TL}^{+\dagger}(p,s)b_{TL}^{+}(p,s) - d_{TL}^{+}(p,s)d_{TL}^{+\dagger}(p,s) - b_{TR}^{+\dagger}(p,s)b_{TR}^{+}(p,s) +$$
$$+ d_{TR}^{+}(p,s)d_{TR}^{+\dagger}(p,s)]$$

$$= \sum_{\pm s} \int d^2p dp^+ p^- [b_{TL}^{+\dagger}(p,s) b_{TL}^{+}(p,s) + d_{TL}^{+\dagger}(p,s) d_{TL}^{+}(p,s) - b_{TR}^{+\dagger}(p,s) b_{TR}^{+}(p,s) - - d_{TR}^{+\dagger}(p,s) d_{TR}^{+}(p,s)]$$

up to the usual infinite constants due to left-handed operator rearrangement and right-handed operator rearrangement that are discarded. The above equation is the basis for our particle interpretation of tachyon creation and annihilation operators based on Dirac's hole theory. Dirac hole theory, as applied in light-front coordinates, assumes all negative p^- ("energy") states are filled.

4.2.4.1 Left-Handed Tachyon Creation and Annihilation Operators

1. We identify $b_{TL}^{+\dagger}(p,s)$ and $d_{TL}^{+}(p,s)$ as creation operators for left-handed tachyons. $b_{TL}^{+\dagger}(p,s)$ creates a positive p^- ("energy") state and $d_{TL}^{+}(p,s)$ creates a negative p^- ("energy") state.

2. $b_{TL}^{+}(p,s)$ and $d_{TL}^{+\dagger}(p,s)$ are the corresponding annihilation operators for left-handed tachyons. $b_{TL}^{+}(p,s)$ annihilates a positive p^- ("energy") state and $d_{TL}^{+\dagger}(p,s)$ annihilates a negative p^- ("energy") state.

3. We assume Dirac hole theory holds for the left-handed tachyon vacuum with all negative energy states filled. There is no tachyon energy gap as there is for Dirac fermions. There is also the problem that the left-handed tachyon vacuum is not invariant under ordinary Lorentz transformations or Superluminal transformations. *However if we confine ourselves to light-front coordinates for computations no ambiguity can result and the Lorentz covariant quantities that we calculate, such as the S matrix, are well-defined.*

4. Using tachyon hole theory we identify $b_{TL}^{+}(p,s)$ and $d_{TL}^{+\dagger}(p,s)$ as annihilation operators for left-handed tachyons. $b_{TL}^{+}(p,s)$ annihilates a positive p^- ("energy") state and $d_{TL}^{+\dagger}(p,s)$ annihilates a negative p^- ("energy") state – thus creating a hole in the tachyon sea that we view as the creation of a positive p^- ("energy"), left-handed antitachyon. $d_{TL}^{+}(p,s)$ annihilates a positive p^- ("energy"), left-handed antitachyon.

4.2.4.2 Right-Handed Tachyon Creation and Annihilation Operators

The anti-commutation relations of right-handed tachyon creation and annhilation operators and the right-handed Hamiltonian terms have the "wrong" sign compared to corresponding Dirac operators and left-handed tachyon operators. This situation is completely analogous to the situation of time-like photons in the covariant formulation of quantum Electrodynamics.[39] In the case of time-like photons it was possible to introduce an indefinite

[39] Bogoliubov (1959) pp. 130-136.

metric (Gupta-Bleuler formulation), and then to use the subsidiary condition $\partial A^{\nu}/\partial x^{\nu} = 0$ to reduce the dynamics of QED to the transverse components. Thus the time-like photons were intermediate artifacts needed to have a manifestly covariant formulation while QED observables depended solely on the transverse components of the electromagnetic field.

In the present case of free tachyons, and in leptonic ElectroWeak Theory, there is no evident "subsidiary condition" to eliminate the right-handed tachyon fields. But since the only manner in which the right-handed leptonic tachyon fields[40] interact is through mass terms, which can be easily 'integrated out", right-handed leptonic tachyon fields are removed from the observable part of the leptonic ElectroWeak Theory by their "lack of interaction" with left-handed fields.

In the case of quark ElectroWeak Theory right-handed tachyon quark fields have charge (−1/3) and thus experience an electromagnetic interaction as well as a Z interaction. However, *since quarks are totally confined, right-handed tachyon quarks will not be able to continuously emit photons or Z's due to energy conservation and their confinement to bound states of fixed positive energy.*

Thus right-handed tachyons are analogous to time-like photons – necessary theoretically but prevented from causing a negative energy disaster by the forms of their interactions.

4.2.4.3 Comments

As a result of these considerations we find Parity Violation with left-handed fermions dominating the ElectroWeak interactions.[41] *Thus the Complex Lorentz group, when used to generate fermion species and tachyons in particular, leads through canonical quantization of fermion fields directly to Parity Violation.*

---

[40] The tachyon fields are provisionally assumed to be neutrino fields in the leptonic sector, and d, s and b quarks in the quark sector.

[41] We note that sign conventions are relevant. We use the metric $[\eta_{\mu\nu}] = \mathrm{diag}(1, -1, -1, -1)$ throughout this book.

4.2.5 Complex Lorentz transformations and the Reality group

The Real Lorentz group transforms between coordinate systems that are in relative motion at a velocity below the speed of light. Consequently it transforms real-valued coordinate systems to real-valued coordinate systems. The Complex Lorentz group includes the Real Lorentz group but also includes transformations that transform real-valued coordinate systems into complex-valued coordinate systems. A subset of these transformations transform between coordinate systems at a velocity whose magnitude is below the speed of light. Another subset of transformations transform between coordinate systems at a relative velocity whose magnitude is above the speed of light.[42] These transformations correspond to faster than light motion and provide the boosts mentioned above that generate tachyonic fermions.

We showed in earlier books such as Blaha (2017b) that faster than light physics is consistent and physically acceptable – even to the point of deriving faster-than-light Thermodynamics from the Maxwell-Boltzmann distribution including the law of increasing Entropy.

In this subsection[43] we will outline the physical basis of the known Standard Model symmetry group SU(3)⊗SU(2)⊗U(1) and show that it should generalize to

$$SU(3)\otimes SU(2)\otimes U(1)\otimes SU(2)\otimes U(1)$$

where the extra SU(2)⊗U(1) factor describes the Dark Matter ElectroWeak interactions.[44] The normal matter SU(2)⊗U(1) symmetry emerges from the consideration of boosts at real-valued velocities greater than the speed of light. The Dark Matter sector SU(2)⊗U(1) symmetry emerges from the consideration of boosts at complex-valued velocities less than or greater than the speed of light.

Before beginning the discussion of these cases we note that a local U(4) transformation can change any complex-valued coordinate system to a real-valued coordinate system in 4-dimensional flat space-time. However because of the peculiar nature of the Lorentz group, and the defining relation of the Real and Complex Lorentz groups:

$$\Lambda(\mathbf{v}, \boldsymbol{\theta})^T G \Lambda(\mathbf{v}, \boldsymbol{\theta}) = G$$

---

[42] This was not noted by the rigorously mathematical derivation of Axiomatic Quantum Field Theory by Streater (2000) and others.

[43] This subsection outlines the detailed discussion presented in chapters 5 – 7 of Blaha (2017b) as well as earlier books such as Blaha (2015a).

[44] The Dark Matter ElectroWeak interactions must be distinct from the normal matter ElectroWeak interactions or Dark Matter would have been found experimentally by now.

where G is the flat space-time metric G = diag(1, –1, –1, –1), and where the superscript ^{T} specifies the transpose of the matrix; it is possible to *physically* specify the origin of each of the subgroups of U(4): SU(2)⊗U(1), SU(3), and 'Dark' SU(2)⊗U(1). (See below.) These subgroups do not commute with each other. The total number of their generators is 16 as is the number of generators of U(4).

Now following Decision Axiom 3.1.3.6 'Nature Tends to Repeat Successful Strategies' we assume that the fundamental internal symmetry group of elementary particles is analogous to the set of subgroups of U(4) that transform complex-valued coordinate systems into real-valued coordinate systems. However we will construct it by creating a product of the subgroup types with the assumption that the subgroups, which are now internal symmetry subgroups, commute with each other. Thus the choice of the internal symmetry Reality group as the tensor product: SU(3)⊗SU(2)⊗U(1)⊗SU(2)⊗U(1). We denote it as R and call it the Reality group due to its origin in Complex Lorentz group space-time transformations:

$$R = SU(3)\otimes SU(2)\otimes U(1)\otimes SU(2)\otimes U(1)$$

Each of the factors has an associated set of Yang-Mills vector boson fields that become particle interactions. Thus local gauge theories.

We will now show that Complex Lorentz transformations can be classified in a simple manner based on the U(4) subgroup factors above.[45] This classification reinforces our choice of R as the internal symmetry group due to its close correspondence with fermion species described earlier and discussed further in subsection 4.2.5.3 below.

It is important to note that the five factors above exhaust the possible Reality group transformations of coordinates. Thus by analogy (Decision Axiom 6) R is the complete set of interactions associated with elementary particles due to the Complex Lorentz group.[46]

4.2.5.1 Boosts at Real-valued Velocities Greater Than the Speed of Light

In chapter 5 of Blaha (2017b) we showed that Complex Lorentz boosts at real-valued velocities with magnitude greater than the speed of light transform a rest frame coordinate system to a new complex-valued coordinate system. In general an SU(2)⊗U(1) transformation is needed to transform the target complex-valued coordinate system to a real-valued coordinate system. We can symbolize these transformations with

$$\text{Real-valued Coordinate System} \rightarrow R_{SU(2)\otimes U(1)}\Lambda(|\mathbf{v}| > \mathbf{c}) \rightarrow \text{Real-valued Coordinate System}$$

[45] See Blaha (2017b) as well as Blaha (2015a) and other earlier books by the author.

[46] Later we show that there are four layers of fermions, each with its own separate R Reality group set of transformations.

where $\Lambda(|\mathbf{v}| > \mathbf{c})$ is a Complex Lorentz transformation with real-valued relative velocity $|v| > c$, and $R_{SU(2)\otimes U(1)}$ is an $SU(2)\otimes U(1)$ transformation from complex-valued coordinates to real-valued coordinates.

The Complex Lorentz transformation $\Lambda(|\mathbf{v}| > \mathbf{c})$ when applied to a fermion at rest transforms it into a tachyon with real-valued energy. (See chapter 3 of Blaha (2017b).) The tachyons of this type are those of the neutral lepton species and of the Dark neutral lepton species. (See section 4.2.3 above.)

4.2.5.2 Boosts at Complex-valued Velocities

There are two types of boosts of this kind: those whose magnitude of velocity exceeds the speed of light yielding tachyons, and those whose magnitude of velocity is below the speed of light yielding a non-tachyon fermion. (See chapters 6 and 7 of Blaha (2017b) for details.) The forms of the tachyons of these varieties are described above in section 4.2.3. The fermions generated by these boosts are the up and down type normal and Dark quarks.

4.2.5.3 Mapping of the Boosted Fermions to Normal and Dark Fermion Species

We have seen the four types of fermion species, and the internal symmetries and interactions of particles, emerge in this section from the Complex Lorentz group. We can summarize the map from our theory to the real world in the following way:

Normal Matter Fermions

Dirac fermions – Charged Leptons – Fields generated by Real Lorentz boosts – real $v < c$
Tachyon fermions – Neutrinos – Fields generated by Real Lorentz boosts – real $v > c$
Complexon fermions – Up-Type Quark Triplets – Fields generated by Lorentz boosts – complex $v < c$
Tachyon Complexon fermions – Down-Type Quark Triplets – From Lorentz boosts – complex $v > c$

Dark Matter Fermions

Dirac fermions – Dark Charged Leptons – Fields generated by Real Lorentz boosts – real $v < c$
Tachyon fermions – Dark Neutrinos – Fields generated by Real Lorentz boosts – real $v > c$
Complexon fermions – Dark Up-Type Quark Singlets – Fields generated by Lorentz boosts – complex $v < c$
Tachyon Complexon fermions – Dark Down-Type Quark Singlets – From Lorentz boosts – complex $v > c$

Normal Matter Gauge Bosons

$SU(2)\otimes U(1)$ - real space-time coordinates – not complexon coordinates
$SU(3)$ - complex space-time coordinates – complexon coordinates

Dark Matter Gauge Bosons

$SU(2)\otimes U(1)$ - real space-time coordinates – not complexon coordinates

4.2.5.4 Comments

1. In making the choice of coordinates for gauge fields we chose complexon coordinates for SU(3) because quarks have complexon coordinates. Real and imaginary spatial momenta are separately conserved. Thus significant dynamics requires complexon SU(3) fields.

2. We assumed the analogous form of subgroups in the coordinate and internal symmetry reality subgroups because of the near match of subgroups (Dark Matter presumably has the other SU(2)⊗U(1) group) based on Ockham's Razor and Decision Axiom 6 (near match of forms).

3. We arbitrarily chose to match the coordinate dependence of normal and Dark leptons as real-valued and not complexon based on Ockham's Razor and Decision Axiom 6.

4. We chose the Dark quark fields to have complexon coordinates in analogy with normal quark fields. Since Dark Matter does not exhibit Strong Interactions with normal matter we chose the Dark quarks to be SU(3) singlets.

5. We chose the Dark SU(2)⊗U(1) fields to have real-valued coordinates in analogy with normal SU(2)⊗U(1) fields. (Decision Axiom 6 and Ockham's Razor)

6. We chose normal and Dark, up-type quarks to be complexon, and down-type quarks to be tachyonic complexon, due to the close connection of down-type quarks with parity violation and thus a tachyonic nature.
7. Our formulation closely depends on Complex Lorentz group features. We view this as an arguement for the correctness of our theory.

4.2.6 Two-Tier Coordinates

Originally Two-Tier coordinates were developed by this author to remove infinities that appear in perturbation theory calculations. We showed that the quantum smeared coordinates of Two-Tier Quantum Field Theory succeeded in removing all ultra-violet infinities in perturbation theory including the fermion triangle infinities. Remarkably the high precision, low energy[47] predictions of QED remained true in Two-Tier QED and thus remained consistent with experiment to a hitherto unsurpassed level of accuracy. 'Low' energy predictions in other quantum field theories also remained unchanged. At high energies, Two-Tier perturbation theory results are finite and consequently all ultra-violet infinities, to any order in perturbation theory, in *any number of space-time dimensions* were eliminated.

In addition to removing perturbation theory infinities Two-Tier coordinates enable us to define finite theories of Quantum Gravity and 'non-renormalizable' quantum field theories based on polynomial lagrangians, to tame vacuum fluctuations, to eliminate infinities associated with the Big Bang, and to generate the explosive growth of the universe in its role as Dark Energy.[48] In this book, in chapters 1 and 2 we show that the Two-Tier field Y^μ, which has the features of the free electromagnetic field, originates in Quantum Logic. Thus Two-Tier Quantum Field Theory is established on the most fundamental level.

4.2.6.1 Two-Tier Features in 4-Dimensional Space-Time

Two-Tier Quantum Field Theory,[49] which was based on a new method in the Calculus of Variations, uses two 'layers' of fields to introduce quantum coordinates. We shall consider this technique for the specific case of a massless vector field $V^i(y)$ analogous to the electromagnetic field.

In 4-dimensional space-time the massless vector field has the form $Y^\mu(y)$ where the index μ ranges from 0 through 3. The X^μ coordinate system, where it appears, has a c-number real part and a q-number imaginary part. Thus particle fields which are normally defined on four-dimensional real space-time will now be defined on a complex four-dimensional space-time where four imaginary dimensions will appear as *Quantum Dimensions* embodied in a vector quantum field $Y^\mu(y)$.

$$X^\mu(y) = y^\mu + i\, Y^\mu(y)/M_c^2$$

where M_c is an extremely large mass of the order of the Planck mass or perhaps much larger.

---

[47] Relative to a mass scale that was perhaps of the order of the Planck mass.

[48] See Blaha (2017b) and earlier books for details. This section is basically a summary of some features.

[49] See Blaha (2005a), and Blaha (2002), for discussions of this new method to eliminate infinities in quantum field theory calculations.

The $Y^{\mu}(y)$ field is a function of the subspace y coordinates. The real part of the space-time dimensions will be taken to be the space of real-valued y coordinates.[50]

The imaginary part of space-time coordinates is the a massless $Y^{\mu}(y)$ vector quantum field that is suppressed further by a very large mass scale – perhaps of the order of the Planck mass – that reduces the imaginary Quantum Dimensions to the infinitesimal except at large momenta. The effects of Quantum Dimensions only become appreciable in quantum field theory at energies of the order of M_c. At these energies exponential Gaussian factors in each particle (and ghost) propagator that is generated by the Quantum Dimensions serves to make perturbation theory calculations ultra-violet finite – including calculations in Quantum Gravity.

The formalism introduces a new form of interaction that does not have the form of the simple polynomial interactions that have hitherto dominated quantum field theories. This form of interaction takes place via the composition of quantum fields and can be called a *Dimensional Interaction* or an *Interdimensional Interaction* since it affects particle behavior through Quantum Dimensions.

The basic ansatz of the Two-Tier formalism is to replace every appearance of a coordinate x in a quantum field with the variable

$$x^{\mu} \rightarrow X^{\mu} = (y^0, \mathbf{y} + \mathbf{Y}(y^0, \mathbf{y})/M_c^2)$$

where $\mathbf{Y}(y^0, \mathbf{y})$ is the spatial part of a free massless vector field with features that are identical to the free QED field in the Radiation gauge.

Then one finds that the momentum space free field Feynman propagators G(k) of all particles acquires a Gaussian factor exp(h(k)):

$$G(k) \rightarrow G(k)\,\exp(h(k))$$

so that all perturbation theory diagrams are finite. The result is finite perturbative results for all calculations to any order in perturbation theory. Blaha (2005a) shows that Two-Tier theories are finite, Poincare covariant, and unitary. (See Blaha (2005a), part 5, for a complete discussion.)

4.2.6.2 Simple Two-Tier X^{μ} Formalism

In this subsection we will describe the basic Two-Tier formalism. Taking the lagrangian described in Blaha (2005a):[51]

$$\mathscr{L}(y) = \mathscr{L}_F(X^{\mu}(y))J + \mathscr{L}_C(X^{\mu}(y), \partial X^{\mu}(y)/\partial y^{\nu}, y)$$

where

[50] In a deeper theory the real part might also be a quantum field that undergoes a condensation to generate c-number coordinates. We will not consider this possibility in this book.

[51] Eq. 7.1.

$$X^\mu(y) = y^\mu + i\, Y^\mu(y)/M_c^2$$

with M_c being a large mass scale, $Y_\mu(y)$ a vector quantum field, and where J is the absolute value of the Jacobian of the transformation from X to y coordinates:

$$J = |\partial(X)/\partial(y)|$$

The lagrangian term $\mathscr{L}_C$ is

$$\mathscr{L}_C = +\tfrac{1}{4}\, M_c^4 F^{\mu\nu} F_{\mu\nu}$$

with

$$F_{\mu\nu} = \partial X_\mu/\partial y^\nu - \partial X_\nu/\partial y^\mu$$
$$\equiv i\,(\partial Y_\mu/\partial y^\nu - \partial Y_\nu/\partial y^\mu)/M_c^2$$

The lagrangian term $\mathscr{L}_F\,(X^\mu(y))$ contains the terms for scalar, fermion and other gauge terms in general. The sign in $\mathscr{L}_C$ is not negative – contrary to the conventional electromagnetic Lagrangian. The reason for this difference is that the quantum field part of X^μ is imaginary. Thus $\mathscr{L}_C$ ends up having the correct sign after taking account of the factor of i in the field strength $F_{\mu\nu}$.

Defining

$$F_{Y\mu\nu} = (\partial Y_\mu/\partial y^\nu - \partial Y_\nu/\partial y^\mu)$$

we see the Lagrangian assumes the form of the conventional electromagnetic Lagrangian:

$$\mathscr{L}_C = -\tfrac{1}{4}\, F_Y^{\mu\nu} F_{Y\mu\nu}$$

The action of this theory has the form

$$I = \int d^4y\, \mathscr{L}(y)$$

4.2.6.3 Y^μ Gauge

The gauge invariance of the Lagrangian allows us to choose a convenient gauge. The gauge invariance of the full Lagrangian

$$\mathscr{L}_s = L_F(\phi(X), \partial\phi/\partial X^\mu)\, J + \mathscr{L}_C(X^\mu(y), \partial X^\mu(y)/\partial y^\nu)$$

is based on the standard gauge invariance of $\mathscr{L}_C$, and the gauge invariance of $J\mathscr{L}_F$ in the form of translational invariance

$$X^\mu(y) \to X^\mu(y) + \delta X^\mu(y)$$

for the special case of a translation of X with the form of a gauge transformation:

$$\delta X^{\mu}(y) = \partial\Lambda(y)/\partial y_{\mu}$$

In this case we find

$$\int d^4y\ \Lambda(y)\ \partial\,[\ J\ \partial/\partial X^{\mu}\ \mathscr{T}_{F\mu\nu}\ ]/\partial y_{\nu} = 0$$

after a partial integration and so we have the differential conservation law:

$$\partial\,[\ J\ \partial\mathscr{T}_{F\mu\nu}/\partial X^{\mu}]/\partial y_{\nu} = 0$$

since $\Lambda(y)$ is arbitrary. This conservation law is trivially obeyed:

$$\partial\mathscr{T}_{F\mu\nu}/\partial X^{\mu} = 0$$

Thus translational invariance in the $\mathscr{L}_F$ sector together with standard gauge invariance in the $\mathscr{L}_C$ sector automatically guarantees Y field gauge invariance of the total Lagrangian. We use the separate invariance of each term of

$$L = \int d^4y\ [\mathscr{L}_F\, J + \mathscr{L}_C\,] = \int d^4X\ \mathscr{L}_F + \int d^4y\ \mathscr{L}_C = L_F + L_C$$

under a constant translation $X^{\mu} \to X^{\mu} + \delta X^{\mu}$ where δX^{μ} is constant. Then we consider a position dependent translation/gauge transformation, which taken together with the above equation, establishes the invariance under the position dependent translation/gauge transformation.

An alternate approach that leads to the same result is to start with the particle part of the Lagrangian $\mathscr{L}_F$ rewritten to be invariant under general coordinate transformations, as it must, when we generalize to include General Relativity. Since position dependent translations are a form of general coordinate transformation the full theory must be invariant under position dependent translations due to invariance under general coordinate transformations.

Having established invariance under gauge transformations we now choose to use the most convenient gauge – the radiation gauge[52]:

$$\partial Y^i/\partial y^i = 0$$

[52] It is also possible to quantize using an indefinite metric that preserves manifest Lorentz covariance as was done by Gupta and Bleuler for the electromagnetic field. We will use the Gupta-Bleuler approach later to establish covariance under special relativity later. Now we opt for manifest positivity and use the radiation gauge.

where i = 1, 2, 3, which, in the absence of external sources, allows us to set

$$Y^0 = 0$$

since Y^0 does not have a canonically conjugate momentum. A conventional treatment leads to the equal time commutation relations:

$$[Y^\mu(\mathbf{y}, y^0), Y^\nu(\mathbf{y'}, y^0)] = [\pi^\mu(\mathbf{y}, y^0), \pi^\nu(\mathbf{y'}, y^0)] = 0$$

$$[\pi^j(\mathbf{y}, y^0), Y_k(\mathbf{y'}, y^0)] = -i\,\delta^{tr}{}_{jk}(\mathbf{y} - \mathbf{y'})$$

(Note the locations of the j indexes above introduce a minus sign.) where

$$\pi^k = \partial\mathscr{L}_C \,/\, \partial Y_k'$$
$$\pi^0 = 0$$

$$\delta^{tr}{}_{jk}(\mathbf{y} - \mathbf{y'}) = \int d^3k\, e^{i\,\mathbf{k}\cdot(\mathbf{y} - \mathbf{y'})}(\delta_{jk} - k_j k_k/\mathbf{k}^2)/(2\pi)^3$$

$$Y_k' = \partial Y_k/\partial y^0$$

The Radiation gauge reveals the two degrees of freedom that are present in the vector potential. The Fourier expansion of the vector potential is:

$$Y^i(y) = \int d^3k\, N_0(k) \sum_{\lambda=1}^{2} \varepsilon^i(k, \lambda)[a(k,\lambda)\, e^{-ik\cdot y} + a^\dagger(k,\lambda)\, e^{ik\cdot y}]$$

where

$$N_0(k) = [(2\pi)^3 2\omega_k]^{-1/2}$$

and (since m = 0)

$$\omega_k = (\mathbf{k}^2)^{1/2} = k^0$$

with $\vec{\epsilon}(k, \lambda)$ being the polarization unit vectors for $\lambda = 1{,}2$ and $k^\mu k_\mu = 0$.

The further development of this theory is described in Part 3 of Blaha (2005a).

4.2.6.4 Scalar Field Quantization Using X^μ

We will begin by considering the case of a scalar quantum field theory. We assume a real underlying y subspace. Since X^μ is a set of coordinates, we choose to define a scalar field ϕ

as a function of X^μ, which, in turn, is a function of the y^ν coordinates. We will provisionally second quantize ϕ treating X^μ as c-number coordinates using a conventional approach.[53]

We assume a Lagrangian, with the momentum conjugate to ϕ:

$$\pi_\phi = \partial L_F / \partial \phi' \equiv \partial L_F / \partial(\partial\phi / \partial X^0)$$

Following the canonical quantization procedure, π and ϕ become hermitian operators with equal time ($X^0 = X^{0\prime}$) commutation rules:

$$[\phi(X), \phi(X')] = [\pi_\phi(X), \pi_\phi(X')] = 0$$

$$[\pi_\phi(X), \phi(X')] = -i\,\delta^3(\mathbf{X} - \mathbf{X'})$$

The standard Fourier expansion of the solution to the Klein-Gordon equation is:

$$\phi(X) = \int d^3p\, N_m(p)\, [a(p)\, e^{-ip\cdot X} + a^\dagger(p)\, e^{ip\cdot X}]$$

where

$$N_m(p) = [(2\pi)^3 2\omega_p]^{-1/2}$$

and

$$\omega_p = (\mathbf{p}^2 + m^2)^{1/2}$$

The commutation relations of the Fourier coefficient operators are:

$$[a(p), a^\dagger(p')] = \delta^3(\mathbf{p} - \mathbf{p'})$$

$$[a^\dagger(p), a^\dagger(p')] = [a(p), a(p')] = 0$$

The reader will recognize the quantization procedure is formally identical to the standard canonical quantization procedure of a free scalar quantum field.

In the case of spin ½, spin 1 and spin 2 fields the standard quantization procedure *in terms of the X coordinate system* can also be followed in a way similar to the procedure in standard texts.

---

[53] Some texts are: Bogoliubov, N. N., Shirkov, D. V., *Introduction to the Theory of Quantized Fields* (Wiley-Interscience Publishers Inc., New York, 1959); Bjorken, J. D., Drell, S. D., *Relativistic Quantum Fields* (McGraw-Hill, New York, 1965); Huang, K., *Quarks, Leptons & Gauge Fields Second Edition* (World Scientific, River Edge, NJ, 1992); Kaku, M., *Quantum Field Theory* (Oxford University Press, New York, 1993); Weinberg, S., *The Quantum Theory of Fields* (Cambridge University Press, New York, 1995).

4.2.6.5 Scalar Feynman Propagators

The momentum space free field Feynman propagators G…(k) of all particles and ghosts in all Two-Tier Quantum Field Theories acquires a Gaussian factor exp(h(k)):

$$G\ldots(k) \rightarrow G\ldots(k)\ \exp(h(k))$$

so that all perturbation theory diagrams are finite. The result is a finite perturbative result in all calculations to any order in perturbation theory. Blaha (2005a) shows that Two-Tier theories are finite, Poincare covariant, and unitary.

An example of the Two-Tier effect on propagators is the case of the Two-Tier photon propagator. The Two-Tier photon propagator[54] is:

$$iD_F^{TT}(y_1 - y_2)_{\mu\nu} = -i \int \frac{d^4p\ e^{-ip\cdot z}\ g_{\mu\nu}\ R(\mathbf{p}, z)}{(2\pi)^4\ (p^2 + i\varepsilon)}$$

(since the imaginary parts can be taken to be zero: $y_{1i}^{\mu} - y_{2i}^{\mu} = 0$) where

$$z^{\mu} = y_{1r}^{\mu} - y_{2r}^{\mu}$$

$$R(\mathbf{p}, \mathbf{z}) = \exp[-p^i p^j \Delta_{Tij}(z)/M_c^4]$$

$$= \exp\{ -\mathbf{p}^2[A(v) + B(v)\cos^2\theta] \,/\, [4\pi^2 M_c^4 |\mathbf{z}|^2$$

with i, j = 1, 2, 3, and with $\Delta_{Tij}(z)$ being the commutator of the positive frequency part $Y^+{}_k(y)$ and the negative frequency part $Y^-{}_k(y)$ of $Y_k(y)$:

$$\Delta_{Tij}(z) = [Y^+{}_j(y_{1r}), Y^-{}_k(y_{2r})] = \int d^3k\ e^{ik\cdot(y_{1r} - y_{2r})}\ (\delta_{jk} - k_j k_k/\mathbf{k}^2)/[(2\pi)^3 2\omega_k]$$

and

$$v = |z^0|/|\mathbf{z}|$$
$$A(v) = (1 - v^2)^{-1} + .5v\ \ln[(v - 1)/(v + 1)]$$
$$B(v) = v^2(1 - v^2)^{-1} - 1.5v\ \ln[(v - 1)/(v + 1)]$$
$$\mathbf{p}\cdot\mathbf{z} = |\mathbf{p}|\ |\mathbf{z}|\ \cos\theta$$

with $|\mathbf{p}|$ denoting the length of a spatial vector **p**, $|\mathbf{z}|$ denoting the length of a spatial vector **z**, and with $|z^0|$ being the absolute value of z^0.

---

[54] Blaha (2005a).

The gaussian factors R(**p**, z) which appear in all Two-Tier propagators damp the large momentum behavior of all perturbation theory integrals producing a completely finite perturbation theory and yet give the usual results of perturbation theory at energies that are small compared to the mass scale M_c.

4.2.6.6 String-like Substructure of the Theory

Two-tier Quantum field Theory endows each particle with an extended structure that resembles the extended structure seen in bosonic string and Superstring theories.[55] For example, Bailin (1994) use the operator[56]

$$V_\Lambda(k) = \int d^2\sigma \sqrt{-h}\, W_\Lambda(\tau, \sigma)\, e^{-ik\cdot X}$$

where X^μ is a quantized fourier expansion of the string fields (see eq. 7.22 of Bailin (1994)).

We note our X^μ coordinate-field has two transverse degrees of freedom due to gauge invariance, which also invites comparison to the bosonic string. A point of difference is that we have a well-defined quantum field theoretic formulation in conventional space-time that has the Standard Model as its "large distance" behavior thus introducing a note of reality that is not apparent in Superstring theories. We see that the interacting quantum field theories based on this approach also have good, finite, short distance behavior just as string theories.

The scalar, and other particles', Feynman propagators can be viewed as describing the propagation of a particle cloaked (accompanied) by a cloud of Y particles (which generates the R(**p**, $y_1 - y_2$) factor in the above propagator). If we examine the fourier transform of R(p, z) we see:

$$(2\pi)^4 R(\mathbf{p}, q) = \int d^4z\, e^{iq\cdot z}\, R(\mathbf{p}, z) = \int d^4z\, e^{iq\cdot z} \exp[-p^i p^j \Delta_{Tij}(z)/M_c^4]$$

and we find

$$R(\mathbf{p},q) = \sum_{n=0}^{\infty} [i(2\pi M_c)^4]^{-n} (n!)^{-1} \prod_{j=1}^{n} [\int d^4k_j\, \theta(k_j^0)(\mathbf{p}^2 - (\mathbf{p}\cdot\mathbf{k}_j)^2/\mathbf{k}_j^2)/(k_j^2 + i\varepsilon)]\, \delta^4(q - \sum k_r)$$

which can be interpreted as a "cloud" of Y particles dressing the "bare" particle propagator. (The apparent divergences for R(p, q) are an artifact of the expansion and the subsequent fourier transformation. They are not present in the R(**p**, $y_1 - y_2$) factor in the propagator. See Fig. 4.1 for the Feynman diagram of the Two-Tier 'cloaked' propagator as compared to the normal scalar particle Feynman propagator. The Two-Tier Feynman propagator is basically a conventional scalar propagator that is modified by coherent Y particle emission.[57]

---

[55] Chapter 2 discussed the string interpretation of Two-Tier Quantum Field Theory in more detail.

[56] D. Bailin and A. Love, *Supersymmetric Gauge Field Theory and String Theory* (Institute of Physics Publishing, Philadelphia, PA, 1994) page 272.

[57] T. W. B. Kibble, Phys. Rev. **173**, 1527 (1968) and references therein. In particular see p. 1532 of Kibble's paper.

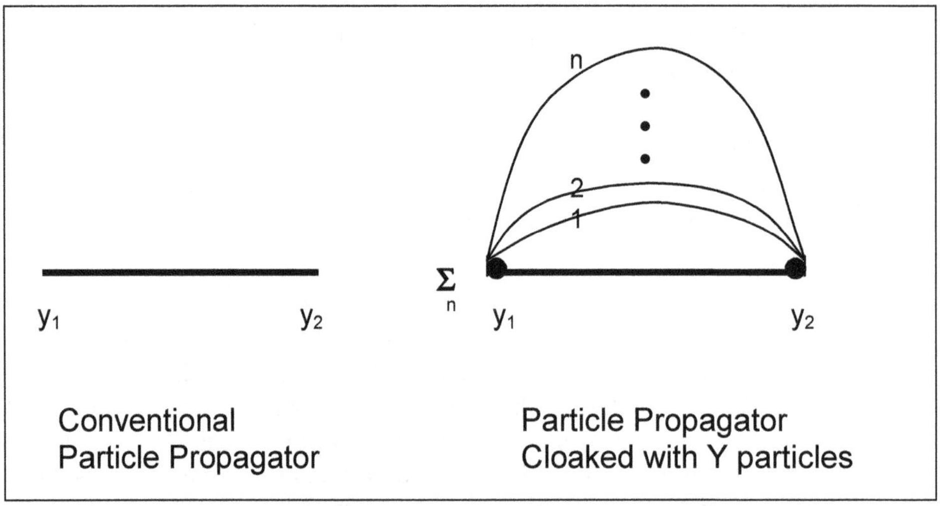

Figure 4.1. Feynman diagram for conventional and the nth diagram of a cloaked Two-Tier propagator.

We note that R(p, q) satisfies the convolution theorem:

$$\int d^4k\, R(\mathbf{p}, k)\, R(\mathbf{p}, q - k) = [R(\mathbf{p}, q)]^2$$

or

$$(2\pi)^4 \int d^4z\, e^{iq\cdot z}\, R(\mathbf{p}, z)\, R(\mathbf{p}, z) = [\int d^4z\, e^{iq\cdot z}\, R(\mathbf{p}, z)\,]^2$$

The proof follows from the Binomial theorem.

4.2.6.7 Two-Tier Complexon Quantum Fields

In the case of the Complexon Standard Model we will need two variables X_r^{μ} and X_i^{μ} since we have complex spatial 3-coordinates. We define them similarly to the previous case:

$$X_r^{\mu}(y_r) = y_r^{\mu} + i\, Y_r^{\mu}(y_r)/M_c^2$$

$$X_i^{\mu}(y_i) = y_i^{\mu} + i\, Y_i^{\mu}(y_i)/M_c^2$$

where we choose the same mass scale for both the "real" and "imaginary" variables. The Two-Tier, single generation, version of the Complexon Standard Model then has an action of the form

$$I_{CSMtt} = \int dy^0 d^3y_r d^3y_i \,(\mathscr{L}_{CSM}(X_r^\mu(y_r), \mathbf{X}_i^k(y_i))J_2)|_{y_i^0 = 0,\, Y_r^0 = Y_i^0 = 0} +$$

$$+ \int dy_r^0 d^3y_r \,\mathscr{L}_C(X_r^\mu(y_r), \partial X_r^\mu(y_r)/\partial y_r^\nu, y_r) +$$

$$+ \int dy_i^0 d^3y_i \,\mathscr{L}_C(X_i^\mu(y_i), \partial X_i^\mu(y_i)/\partial y_i^\nu, y_i)$$

where the replacements

$$x^\mu \equiv x_r^\mu \rightarrow X_r^\mu(y_r)$$

$$x_i^k \rightarrow X_i^k(y_i)$$

for $\mu = 0, 1, 2, 3$ and $k = 1, 2, 3$ are made, followed by defining $y_r^0 = y^0$ and making a Complex Lorentz transformation to a frame where $y_i^0 = 0$. J_2 is the absolute value of the Jacobian of the transformation from (X_r, X_i) to (y_r, y_i) coordinates:

$$J_2 = |\partial(X_r, X_i)/\partial(y_r, y_i)|$$

We also choose gauges where $Y_r^0 = Y_i^0 = 0$. These types of transformations and gauge choices are discussed in detail in Blaha (2005a). The lagrangian terms $\mathscr{L}_C(X_r^\mu(y_r), \partial X_r^\mu(y_r)/\partial y_r^\nu, y_r)$ and $\mathscr{L}_C(X_i^\mu(y_i), \partial X_i^\mu(y_i)/\partial y_i^\nu, y_i)$ have the same form:

$$\mathscr{L}_C = +\tfrac{1}{4} M_c^4 F^{\mu\nu} F_{\mu\nu}$$

with

$$F_{\mu\nu} = \partial X_\mu/\partial y^\nu - \partial X_\nu/\partial y^\mu$$
$$\equiv i\,(\partial Y_\mu/\partial y^\nu - \partial Y_\nu/\partial y^\mu)/M_c^2$$

or defining

$$F_{Y\mu\nu} = (\partial Y_\mu/\partial y^\nu - \partial Y_\nu/\partial y^\mu)$$

we see each lagrangian assumes the form of the conventional electromagnetic Lagrangian:

$$\mathscr{L}_C = -\tfrac{1}{4}\, F_Y^{\mu\nu} F_{Y\mu\nu}$$

The lagrangian is supplemented with the following condition on all complexon fields $\Phi_{\ldots}$:

$$(\partial/\partial X_r^{\,k}(y_r))\,(\partial/\partial X_i^{\,k}(y_i))\Phi\ldots = 0$$

summed over k = 1, 2, 3. Non-complexon fields $\Omega\ldots$ in our left-handed formulation satisfy the subsidiary condition:

$$\{(\partial/\partial X_r^{\,k}(y_r))(\partial/\partial X_i^{\,k}(y_i)) - [(\partial/\partial X_r^{\,k}(y_r))^2(\partial/\partial X_i^{\,m}(y_i))^{\,2}]^{1/2}\}\Omega\ldots = 0$$

summed over k = 1, 2, 3 and over m = 1, 2, 3 separately in each of the two terms.

4.2.6.8 Complexon Feynman Propagator

In the case of complexons, the Two-Tier Feynman propagator differs from the non-complexon case by having an integration over imaginary spatial 3-momenta, a derivative of a delta function embodying the orthogonality of the real and imaginary 3-momenta, and two factors of R(**p**, z): one factor being R($\mathbf{p}_r$, z_r) and the other factor being R($\mathbf{p}_i$, z_i) (where the time components $z_r^{\,0} = z^0$ and $z_i^{\,0} = 0$ since there is only one real time coordinate[58]) thus providing large momentum convergence for both real and imaginary 3-momentum integrations.

For a normal scalar particle the Feynman propagator is:

$$i\Delta_{CTF}(x-y) = \theta(x^+ - y^+)\langle 0|\phi_{CT}(x)\,\phi_{CT}(y)|0\rangle + \theta(y^+ - x^+)\langle 0|\phi_{CT}(y)\phi_{CT}(x)|0\rangle$$
$$= i\int d^4p_r d^3p_i (2\pi)^{-7}\delta'(\mathbf{p}_r\cdot\mathbf{p}_i/m^2)e^{-ip^+(x^- - y^-) - ip^-(x^+ - y^+) + i\mathbf{p}_\perp\cdot(\mathbf{x}_\perp - \mathbf{y}_\perp) - i\mathbf{p}_i\cdot(\mathbf{x}_i - \mathbf{y}_i)}/(p^2 + m^2 + i\varepsilon)$$

in conventional quantum field theory.

In the case of Two-Tier quantum field a scalar *complexon* particle has the the Feynman propagator

$$i\Delta_{CTFtt}(x-y) = i\int d^4p_r d^3p_i (2\pi)^{-7}\delta'(\mathbf{p}_r\cdot\mathbf{p}_i/m^2)\, R(\mathbf{p}_r, z_r)R(\mathbf{p}_i, z_i)\cdot$$
$$\cdot e^{-ip^+(x^- - y^-) - ip^-(x^+ - y^+) + i\mathbf{p}_\perp\cdot(\mathbf{x}_\perp - \mathbf{y}_\perp) - i\mathbf{p}_i\cdot(\mathbf{x}_i - \mathbf{y}_i)}/(p^2 - m^2 + i\varepsilon)$$

where the time components $z_r^{\,0} = z^0$ and $z_i^{\,0} = 0$ since there is only one time coordinate, where R(**p**, z) is given in the previous subsection, and where $p^2. = p^{0\,2} - p_r^{\,2} + p_i^{\,2}$.

Propagators for other types of particles are similarly modified in the Two-Tier formalism (See Blaha 2005a).

---

[58] We can arrange for $z_i^{\,0} = 0$ by making a Complex Lorentz transformation to an inertial frame where z is real.

4.2.6.9 Vacuum Fluctuations

While the expectation value of a *conventional* free scalar field $\phi_{conv}(x)$ is zero in a conventional quantum field theory:

$$<0|\phi_{conv}(x)|0> = 0$$

the vacuum fluctuations of *conventional* scalar quantum field theory are quadratically divergent:

$$<0|\phi_{conv}(x)\phi_{conv}(x)|0> = \int d^3p/[(2\pi)^3 2\omega_p]$$

In "Two-Tier" quantum field theory we find the vacuum expectation value of a free field is zero *and the expectation value of the square of the field is also zero:*

$$<0|\phi(X)\phi(X)|0> = \int d^3p\, e^{-p^ip^j\Delta_{Tij}(0)/Mc^4}/[(2\pi)^3 2\omega_p] = 0$$

since the exponential factor in the integral is $-\infty$. The exponent contains

$$\Delta_{Tij}(z) = \int d^3k\, e^{-ik\cdot z}\, (\delta_{ij} - k_ik_j/\mathbf{k}^2)/[(2\pi)^3 2\omega_k]$$

where "T"is for "Two-Tier". Thus *vacuum fluctuations are zero in Two-Tier quantum field theory.* Correspondingly, we will see that renormalization constants are finite in the Two-Tier versions of QED, Electroweak Theory, the Standard Model and Quantum Gravity. See Blaha (2017b) and references therein for more details.

4.2.610 Time Intervals in General Relativity

Wigner[59] has studied the measurement of time intervals in General Relativity and sees a problem in the measurement of extremely short intervals. According to Wigner, the measurement of a time inteval in a region of space requires the measurement of the length of time required for an event to happen. The measurement requires an accurate clock. But the accuracy of the clock is limited by the energy-time uncertainty relation:

$$\Delta E \Delta t \geq \hbar$$

Thus the uncertainty in the clock's time measurement is related to the uncertainty in the clock's energy which is, in turn, related to the uncertainty in the clock's mass:

[59] E. P. Wigner, Rev. Mod. Phys. **29**, 255 (1957); J. Math. Phys. **2**, 207 (1961).

$$\Delta E = (\Delta m)c^2$$

To obtain "infinite" accuracy the uncertainty (fluctuations) in the clock's mass must be infinite and thus the clock's mass must be infinite. Infinite fluctuations in the clock's mass will produce corresponding infinite fluctuations in the gravitational field.

$$\Delta h \propto \Delta E \qquad \text{(in conventional General Relativity)}$$

As a result the notion of space-time and time intervals (which depend on the geometry through General Relativity) become uncertain. Thus, according to Wigner, and others, the concept of time intervals and space-time points becomes questionable.

The Two-Tier version of Quantum Gravity offers a way out of this dilemma. The gravitational force becomes stronger as one goes to shorter distances (higher energies) down to a distance (up to an energy) whose scale is set by M_c. At shorter distances (higher energies) the gravitational force becomes weaker and declines to zero at zero distance. Thus at very high energy the gravitational field fluctuations (Δh) are at worst inversely proportional to the energy (and probably decline by a higher power of inverse energy.) (The same considerations would apply if one chooses to consider fluctuations in the Riemann-Christoffel symbols.)

$$\Delta h < c_1/E < c_1/(\Delta E) \qquad \text{(in Two-Tier Quantum Gravity)}$$

where c_1 is a constant. Thus Wigner's conclusion does not hold in the Two-Tier version of Quantum Gravity as gravitational fluctuations actually become smaller at energies above a critical energy whose scale is set by M_c.

In fact, combining the above equations we see

$$c_1\Delta t/\Delta h \geq \hbar$$

at sufficiently high energy. Therefore the time uncertainty Δt, and the gravitational field fluctuations Δh, can both decrease while maintaining the energy-time uncertainty relation. *Thus the notion of a space-time point "is saved" in Two-Tier quantum gravity.*

4.2.6.11 Vacuum Fluctuations in the Gravitation Fields

While the expectation value of the free graviton field $h_{\mu\nu\text{conv}}(x)$ (weak field approximation) is zero in a conventional quantum field theoric approach:

$$\langle 0|h_{\mu\nu\text{conv}}(x)|0\rangle = 0$$

the vacuum fluctuations of the *conventional* quantum graviton field is quadratically divergent since

$$<0|h_{\mu\nu\text{conv}}(x)h_{\alpha\beta\text{conv}}(x)|0> = \int d^3p\, b'_{\mu\nu\alpha\beta}(p)/[(2\pi)^3\, 2\omega_p] = \infty$$

where $b'_{\mu\nu\alpha\beta}(p)$ is a rational function of the momentum p.

In "Two-Tier" quantum field theory we find

$$<0|h_{\mu\nu}(X)h_{\alpha\beta}(X)|0> = \int d^3p\, b'_{\mu\nu\alpha\beta}(p)\, e^{-p^ip^j\Delta_{Tij}(0)}/[(2\pi)^3 2\omega_p] = 0$$

since the exponential factor in the integrand is $-\infty$. The exponent contains

$$\Delta_{Tij}(z) = \int d^3k\, e^{-ik\cdot z}(\delta_{ij} - k_ik_j/\mathbf{k}^2)/[(2\pi)^3 2\omega_k]$$

Thus the vacuum fluctuations of $h_{\mu\nu}$ are zero in "Two-Tier" quantum field theory and, correspondingly, the weak field Two-Tier quantization of Quantum Gravity is consistently finite (and weak in perturbation theory calculations.)

4.2.6.12 Two-Tier Features in D-Dimensional Space-Time (such as the Megaverse)

Since a field, quantized in D-dimensional conventional coordinates (D > 4), would lead to divergences in perturbation theory calculations, we can use D-dimensional Two-Tier coordinates to avoid divergences in perturbation theory:

$$Y^i(y) = y^i + i\, Y_u^i(y)/M_u^{D/2}$$

where $Y_u^i(y)$ for i = 1, …, D is a D-dimensional free gauge field and M_u is a mass of the order of the Planck mass or greater. The $Y_u^i(y)$ term adds a quantum field to the D coordinates making them a set of quantum coordinates. Quantum coordinate derivatives are defined by

$$\partial_i = \partial/\partial Y^i(y) = \partial/\partial(y^i - Y_u^i(y)/M_u^{D/2})$$

The use of these coordinates to quantize particle fields leads to a completely finite perturbation theory. We applied them in Blaha (2017b) to create a finite fundamental theory of mater. We applied them to fields in the Megaverse[60] to achieve a finite theory of Megaverse dynamics for elementary particles and universe particles.

The second quantization of a vector gauge field $V^i(y)$ is analogous to the second quantization of the electromagnetic field. The lagrangian density terms for the free $V^i(Y(y))$ fields is

$$\mathscr{L}_{Vu} = -\tfrac{1}{4} F_{Vu}{}^{ij}(Y(y))F_{Vuij}(Y(y))$$

---

[60] Blaha (2017c).

The lagrangian is

$$L_{Vu} = \int d^D y \mathscr{L}_{Vu}(Y(y))$$

with

$$F_{Vuij} = \partial V_i(Y(y))/\partial Y^j(y) - \partial V_j(Y(y))/\partial Y^i(y)$$

where the values of i and j range from 1 to D in this section.

The equal time commutation relations, using the Dth coordinate as the time coordinate, are specified in the usual way:

$$[V^i(Y(\mathbf{y}, y^0)), V^j(Y(\mathbf{y}', y^0))] = [\pi^i(Y(\mathbf{y}, y^0)), \pi^j(Y(\mathbf{y}', y^0))] = 0$$
$$[\pi_j(Y(\mathbf{y}, y^0)), V_k(Y(\mathbf{y}', y^0))] = -i\, \delta^{(D-1)tr}{}_{jk}(Y(\mathbf{y},0) - Y(\mathbf{y}',0))$$

where

$$\pi_u{}^k = \partial \mathscr{L}_{Vu}(V(Y(y)))/\partial V_k'(Y(y))$$
$$\pi_u{}^D = 0$$

for k = 1, … , (D – 1), and

$$\delta^{(D-1)tr}{}_{jk}(\mathbf{y} - \mathbf{y}') = \int d^{(D-1)}k\, e^{i\, \mathbf{k}\cdot(Y(\mathbf{y},0) - Y(\mathbf{y}',0))}\, (\delta_{jk} - k_j k_k/\mathbf{k}^2)/(2\pi)^{D-1}$$

$$V_k'(Y(y)) = \partial V_k(Y(y))/\partial y^{1D}$$

for j, k = 1, 2, … , (D – 1).

If we choose the Radiation gauge for $V_k(Y(y))$:

$$V^D(Y(y)) = 0$$
$$\partial V^j(Y(y))/\partial Y^j(y) = 0$$

for j = 1, 2, … , (D – 1) then (D – 2) degrees of freedom (polarizations) are present in the vector potential.[61] The Fourier expansion of the vector potential $V^i(Y(y))$ is:

$$V^i(Y(y)) = \int d^{(D-1)}k\, N_{0V}(k) \sum_{\lambda=1}^{D-2} \varepsilon^i(k, \lambda)[a_V(k,\lambda) :e^{-ik\cdot Y(y)}: + a_V^\dagger(k,\lambda) :e^{ik\cdot Y(y)}:]$$

for i = 1, … , (D – 2) where

$$N_{0V}(k) = [(2\pi)^{(D-1)} 2\omega_k]^{-1/2}$$

and (since the field is massless)

$$k^D = \omega_k = (\mathbf{k}^2)^{1/2}$$

where k^D is the energy, and where the $\varepsilon^i(k, \lambda)$ are the polarization unit vectors for λ = 1, … , (D – 2) and $k^\mu k_\mu = k^{D\,2} - \mathbf{k}^2 = 0$.

---

[61] Note we use the Radiation gauge for $Y^\mu(y)$ also.

The commutation relations of the Fourier coefficient operators are:

$$[a_V(k,\lambda),\ a_V^{\dagger}(k',\lambda')] = \delta_{\lambda\lambda'}\delta^{D-1}(\mathbf{k} - \mathbf{k'})$$
$$[a_V^{\dagger}(k,\lambda),\ a_V^{\dagger}(k',\lambda')] = [a_V(k,\lambda),\ a_V(k',\lambda')] = 0$$

and the polarization vectors satisfy

$$\sum_{\lambda=1}^{D-2} \varepsilon_i(k,\ \lambda)\varepsilon_j(k,\ \lambda) = (\delta_{ij} - k_i k_j/\mathbf{k}^2)$$

The V^{μ} Feynman propagator is

$$iD_F^{trTT}(y_1 - y_2)_{jk} = <0|T(V_j(Y(y_1))V_k(Y(y_2)))|0>$$

$$= -\,ig_{jk} \int \frac{d^Dk\ e^{-ik\cdot(y_1 - y_2)}\ R(\mathbf{k},\ y_1 - y_2)}{(2\pi)^{16}\ (k^2 + i\varepsilon)}$$

where g_{jk} is the D-dimensional Lorentz metric and where $R(\mathbf{k},\ y_1 - y_2)$ is given by

$$R(\mathbf{k},\ y_1 - y_2) = \exp[\,-k^i k^j \Delta_{Tij}(y_1 - y_2)/M_u^D]$$
$$= \exp\{\,-k^2[A(v) + B(v)\cos^2\theta]\ /\ [(2\pi)^{D-2}M_u^4 z^2]\} \qquad)$$

where k^2 is *the sum of the squares of the D – 1 spatial components* with

$$z^{\mu} = y_1^{\mu} - y_2^{\mu}$$
$$z = |\mathbf{z}| = |\mathbf{y}_1 - \mathbf{y}_2|$$
$$k = |\mathbf{k}|$$
$$v = |z^0|/z$$
$$A(v) = (1 - v^2)^{-1} + .5v\ \ln[(v - 1)/(v + 1)]$$
$$B(v) = v^2(1 - v^2)^{-1} - 1.5v\ \ln[(v - 1)/(v + 1)]$$
$$\mathbf{k}\cdot\mathbf{z} = kz\cos\theta$$

and $|\mathbf{k}|$ denoting the length of a spatial (D – 1)-vector $\mathbf{k}$ while $|z^0|$ is the absolute value of $z^0 \equiv z^D$.

As the above equations indicate, the Gaussian damping factor R(k, z) for *all* large spatial momentum k^j is the same for both the positive and negative frequency parts of the (Two Tier) V Feynman propagator. We are assuming the spatial momentum is real-valued in this discussion. It is also important to note that R(k, z) does not depend on $k^0 = k^D$ (in the V and Y_u

Radiation gauges) and thus the integration over k^0 proceeds in the usual way to produce time-ordered positive and negative frequency parts.

The Gaussian exponential factor in *all* spatial coordinates causes the Feynman propagator to be finite and, together with the Gaussian factor in universe particle propagators, causes all perturbation theory calculations when interactions are introduced to be finite as we have seen in Blaha (2017b).

For small momentum much less than M_u then $R(\mathbf{k}, y_1 - y_2) \rightarrow 1$ and the Feynman propagator is the "normal" propagator of conventional D-dimensional quantum field theory. For large momentum the corresponding potential approaches r^{D-3} in contrast to the electromagnetic Coulomb potential r^{-1}. The V potential is highly non-singular at large energies.

Thus using Two-Tier Quantum Field Theory we can perform perturbation theory caluculations that always yield a finite result.[62] This is not true if conventional Quantum Field is used.

---

[62] In particular, the fermion triangle divergence (anomaly) does not occur in our Two Tier Quantum Field Theory of the fermion sector. Thus there is no requirement for axion-like particles in the Megaverse (or in universes) although the possible existence of this type of particle is not ruled out.

4.2.7 PseudoQuantum Field Theory

PseudoQuantum Field Theory (and its Quantum Mechanics analogue CQMechanics[63]) originates in the need to second quantize in unusual coordinate systems, and in curved space-time coordinate systems. The papers in Appendices A and B provide a detailed introduction to PseudoQuantum Field Theory to which the reader is referred.

In this subsection we point out its advantages in a variety of field theory contexts that are relevant for the Unified SuperStandard Model. The advantages of PseudoQuantum Field Theory are:

1. Quantization in any coordinate system in flat or curved space-times with an invariant definition of asymptotic particle states. An n particle asymptotic state in one coordinate system is a unitarily equivalent n particle asymptotic state in any other coordinate system. Therefore particle number is invariant under change of coordinate system. This is important for the Unified SuperStandard Model in curved space-times. It is also important for quantization in higher dimensional Euclidean spaces such as the Megaverse. The method was developed in the late 1970's by the author to provide a quantization procedure which supports a unique particle interpretation of states in arbitrary non-static space-times where no global timelike coordinate (Killing vector) exists. PseudoQuantum Field Theory which we developed in a series of books[64] also can be formulated in the Megaverse. Thus we can use it in the Megaverse to implement the Higgs Mechanism to generate particle masses and symmetry breaking.

2. PseudoQuantum Field Theory enables one to define Higgs particle dynamics in such a way that a non-zero vacuum expectation value cleanly separates from the quantum field part of the Higgs fields. This technique can be used in symmetry breaking mechanisms, mass generation, and possible generation of coupling constants as vacuum expectation values.

3. It supports the canonical definition of higher derivative field theories through the use of the Ostrogradski bootstrap. See Appendix B where a fourth order theory of the Strong interaction is defined that has color confinement and a linear r potential. The potential

[63] See Blaha (2016f). CQMechanics encompasses both classical mechanics and quantum mechanics, and provides a method of rotating between them. It has applications to transitions between Quantum/Semi-Classical Entanglement, and Quantum/Classical Path Integrals, and Quantum/Classical Chaos.

[64] See Blaha (2017b) for the discussion of the PseudoQuantum field theory formalism for Higgs particles in our Extended Standard Model. See chapter 20 of Blaha (2017b), and earlier books, for a more detailed view than that presented here.

part of this theory was used by the Cornell group to calculate the Charmonium spectrum. (See Blaha (2017b) for details.)

An associated advantage of using PseudoQuantum Field Theory is that it provides for retarded propagators and an Arrow of Time.

4.2.7.1 General Case of PseudoQuantization in Differing Coordinate Systems

Appendix A describes the PseudoQuantization procedure that relates seond quantizations in differing coordinate systems. We can epitomize the general concept in the following short example.

Consider the case of a scalar particle in D space-time dimensions that we second quantize in coordinate system denoted 1 with coordinates x based on a timelike Killing vector

$$\varphi(x) = \sum_{\alpha} [\chi_\alpha(x)A_\alpha + \chi_\alpha^*(x)A_\alpha^\dagger]$$

where the $\chi_\alpha(x)$ are positive frequency with respect to a definition of positive frequency within a universe – following the notation of Appendix A.

Consider now the second quantization of the particle field in a second coordinate system denoted 2 with coordinates y based on a different timelike Megaverse Killing vector

$$\varphi(y) = \sum_{\beta} [\psi_\beta(y)b_\beta + \psi_\beta^*(y)b_\beta^\dagger]$$

where the $\psi_\beta(y)$ are positive frequency with respect to 2's definition of positive frequency.

Comparing above definitions we see the difference in the definition of the coordinates used in the field expansions as well as the implicit difference in the definitions of positive frequency. To relate the quantizations to each other, we must use the relation between the x and y coordinates:

$$y_i = f_i(x)$$

or, in vector form,

$$y = f(x)$$

for i = 1, 2, … , D. Thus

$$\varphi(f(x)) = \sum_{\beta} [\psi_\beta(f(x))b_\beta + \psi_\beta^*(f(x))b_\beta^\dagger]$$

Inverting the above equations to obtain the relation of the fourier coefficient operators we see:

$$A_\alpha = \sum_{\beta} [C_{\alpha\beta}\, b_\beta + C'_{\alpha\beta}\, b_\beta^\dagger]$$

where $C_{\alpha\beta}$ and $C'_{\alpha\beta}$ are c-number functions of α and β:

$$C_{\alpha\beta} = (\chi_\alpha(x), \varphi(f(x))$$
$$C'_{\alpha\beta} = (\chi_\alpha{}^*(x), \varphi(f(x))$$

The above equations imply an N particle state in one coordinate system will appear as a superposition of states of various numbers of particles in the other coordinate system IF the standard quantum field theory formulation is used.

To REMEDY this situation – which we take to be unphysical – we must reformulate quantum field theory using the PseudoQuantum formulation presented Appendix A. The scalar particle case is discussed in Appendix A between eqs. 6 – 31, to which the reader is referred.

The conclusions of that section, and the sections following it, in Appendix A are:

1. One can define corresponding unitarily equivalent particle states in two quantizations with invariant particle numbers.
2. The fourier coefficient operators of the two quantizations are related by Bogoliubov transformations and are unitarily equivalent.
3. The group of the local Bogoliubov transformations is an infinite tensor product of $SU_{1,1}$ groups.
4. The vacua of the particle are invariant under Bogoliubov transformations that relate the the Megaverse and the universe quantizations.
5. Unitarily equivalent perturbation theories of both quantizations can be defined.

We now consider the case of Two-Tier PseudoQuantization, and then turn to various applications of PseudoQuantization.

4.2.7.2 Two-Tier PseudoQuantum Field Theory

The combination of the Two-Tier procedure with the PseudoQuantiztion procedure leads to a somewhat more complicated situation. In principle, both are required for a Unified SuperStandard Model in any coordinate system in flat or curved space-times in any number of dimensions. However their direct combination is both complicated and unphysical.

The main purpose of PseudoQuantization is to have particle number invariance under a change of coordinate system. Two-Tier Field Theory 'cloaks' each particle in infinite 'clouds' of Y^μ quanta as Fig. 4.1 illustrates. We define PseudoQuantization as implementing particle number invariance for 'bare' particles without their clouds of Y^μ quanta. Thus an asymptotic particle state of n particles (neglecting its Y^μ quanta cloud) remains a unitarily equivalent n particle state (neglecting its Y^μ quanta cloud) under a change of coordinate system.

To implement this concept we first define quantizations of a particle in coordinate systems without Two-tier quanta. We then 'dress' the quantizations by replacing the coordinates y^μ in each coordinate system with the corresponding Two-Tier coordinates:

$$y^\mu \rightarrow X^\mu(y) = y^\mu + i\, Y^\mu(y)/M_c^2$$

It appears the most convenient gauge in each coordinate system is the Lorentz gauge:

$$\partial Y^{\mu}/\partial y^{\mu} = 0$$

We now briefly consider the case of a scalar particle PseudoQuantization. This case is considered in more detail in Appendix A. Following Appendix A we must introduce two fields $\varphi_1(y)$ and $\varphi_2(y)$ with the free fields' lagrangian

$$\mathscr{L}(y) = \partial^{\mu}\varphi_1\partial_{\mu}\varphi_2 - \tfrac{1}{2}\,\partial^{\mu}\varphi_1\partial_{\mu}\varphi_1 - m^2\varphi_1\varphi_2 + \tfrac{1}{2}\,m^2\,\varphi_1^{\,2}$$

in a coordinate system with coordinates y. Then following the steps indicated in Appendix A from eq. 7 onward we arrive at a PseudoQuantum formulation in the coordinate system with coordinates y that is unitarily equivalent to that of a different coordinate system defined a similar manner.

From eq. 43 onwards we can replace the c-number coordinates x and y with Two-Tier coordinates of the form

$$X^{\mu}(y) = y^{\mu} + i\,Y^{\mu}(y)/M_c^{\,2}$$

and proceed to calculate propagators and perturbation theory diagrams.

Thus we have a straight-forward procedure to unite the PseudoQuantum formalism with Two-Tier coordinates to obtain finite perturbation theory results with unitary equivalence to quantization in other coordinate systems in both flat and curved space-times.

The use of two fields per particle of PseudoQuantum field theory will be seen to part of the applications consider in the remainder of this subsection. We will put aside the consideration of quantizations in other coordinate systems in what follows to keep the presentation as simple as possible.

4.2.7.3 PseudoQuantum Higgs Scalar Particle Field Theory in D-dimensional Space-Time

4.2.7.3.1 The Enigma of Higgs Particles and the Higgs Mechanism

In our previous work on the Standard Model, and its generalization to The Extended Standard Model described in a series of books entitled *Physics is Logic ...*, we showed that the fermion spectrum results from Complex Special Relativity, the gauge interactions result from the Reality group, the fermion generations result from the Generation group, and the Theory of Everything results from a combination with Complex General Relativity. The Higgs particles and the Higgs Mechanism were inserted to generate particle masses and symmetry breaking effects.

Whence comes Higgs particles? A more fundamental cause has not been suggested until our analysis, which is here presented in section 4.2.8. So the Higgs sector appeared to be an expedient mechanism to insert much needed symmetry breaking and masses into the theory.

There are a number of peculiarities in the implementation of the Higgs Mechanism:

1. First, it is selective in the sense that some gauge fields have associated Higgs particles and utilize the Higgs Mechanism, and some gauge fields do not have associated Higgs particles. In particular, the ElectroWeak gauge fields, the Generation group gauge fields, the Layer group fields, and the complex gravity Species gauge fields have associated Higgs particles. The strong interaction (gluon) gauge fields do not.[65]

2. The Higgs potentials have a quadratic mass term of the "wrong" sign plus a quartic interaction term, which together, generate non-zero vacuum expectation values. They obviously accomplish their goal. But the source of these potentials, and why they have the same form, is unknown. One expects a fundamental principle should be operative here.

3. One can imagine creating a Higgs microscope at some super-accelerator. Using this microscope in the presence of a (classical) condensate could enable the Uncertainty Principle to be violated. This possibility, in the case of a microscope using electromagnetic fields, was the source of a heuristic argument for the need to quantize the electromagnetic field.[66]

4. The formulation of the Higgs Mechanism uses classical fields under the assumption that a path integral formulation justifies their use. While this may be true, the path integral formulation relies on implicit, unstated boundary conditions that obscure the physics of the quantum field theoretic nature of the mechanism. A direct quantum field theoretic study of the Higgs Mechanism is needed and would further elucidate its character.

5. Scalar fields have a cloud hanging over them that spin ½ fields do not. A spin ½ particle cannot transition to negative energy because there is a filled sea of negative energy particles. No additional particles can fall into the sea due to the Pauli Exclusion Principle that forbids two fermions with the same 4-momentum and quantum numbers. In the case of scalar particles the Pauli Exclusion Principle does not apply and so a *filled* negative energy sea of scalar particles is not possible and positive energy scalar particles can transition to negative energy without hindrance.

---

[65] See section 4.2.8 for an explanation.

[66] Heitler (1954) p. 86 provides a good discussion of the need to quantize the electromagnetic field.

This problem has been "resolved" by an appropriate definition of the scalar particle vacuum to exclude transitions to negative energy. But the rationale for the definition is lacking. Dirac was asked about this issue many years ago. He said he had a solution to the problem. However he did not present it – in keeping with his well-known taciturn nature. So the issue remains an open question.

For the above reasons we will show that a more satisfactory method of achieving the goals of mass generation and symmetry breaking exists.[67] This method relies on a larger Fock space that enables the appearance of a vacuum expectation value for Higgs particles to be understood within a truly quantum framework. More importantly, this method is a consequence the PseudoQuantization procedure described above that enables unitarily equivalent quantizations in different coordinate systems. So a profound fundamental justification for our Higgs boson formulation exists. One major consequence of this approach is the appearance of a local Arrow of Time – a concept that has been a subject of interest for over one hundred years. Another consequence is a rationale for ElectroWeak Higgs bosons and for their absence for the strong (gluon) interaction.

4.2.7.3.2 PSEUDOQUANTIZATION OF SCALAR PARTICLES

We now consider the PseudoQuantization[68] of a scalar particle field that will become a Higgs particle with a non-zero vacuum expectation value.[69] We begin by defining two fields that correspond to the scalar particle: $\varphi_1(x)$ and $\varphi_2(x)$.[70] These fields will be assumed to have the equal time commutators

$$[\varphi_i(x), \pi_j(y)] = i(1 - \delta_{ij})\delta^3(\mathbf{x} - \mathbf{y})$$
$$[\varphi_i(x), \varphi_j(y)] = 0$$
$$[\pi_i(x), \pi_j(y)] = 0$$

where δ_{ij} is the Kronecker δ and where $\pi_i(x)$ is the canonically conjugate momentum to $\varphi_i(x)$. The fields $\varphi_1(x)$ and $\pi_1(y)$ will be observable classical fields. The fields $\varphi_2(x)$ and $\pi_2(y)$ will not be observables so that $\varphi_1(x)$ and $\pi_1(y)$ can both be sharp on the set of physical states.

---

67 In the Extended Standard Model of Blaha (2015a) we have shown that the basic particles have a mass, the Landauer mass, so that the theory is symmetry violating from the very start. We have also shown that our Two-Tier formalism for quantum field theories always yields finite results in perturbation theory calculations – making the renormalization approach of t'Hooft and others, which relied on initially massless gauge fields, unnecessary.

68 PseudoQuantization in a D-dimensional space-time is described in Blaha (2017c). This discussion is relevant to PseudoQuantization in the Megaverse, or in other universes.

69 Much of this section appears in Blaha (2016c), and earlier books, as well as in S. Blaha, Phys. Rev. **D17**, 994 (1978). The case of fermion PseudoQuantization is also discussed in Appendix A – S. Blaha, Il Nuovo Cimento **49A**, 35 (1979).

70 The subscripts on the fields are not gauge symmetry indices but simply identifiers distinguishing the fields from each other.

We now specify the lagrangian density for a scalar Klein-Gordon particle:

$$\mathcal{L} = \partial\varphi_1/\partial x_\mu \partial\varphi_2/\partial x^\mu$$

with hamiltonian density

$$\mathcal{H} = \pi_1\ \pi_2\ + \partial\varphi_1/\partial x_i \partial\varphi_2/\partial x^i$$

where i labels spatial coordinates, and $\pi_1 = \partial\varphi_2/\partial t$ and $\pi_2 = \partial\varphi_1/\partial t$. The lagrangian $\mathcal{L}$ is without a potential or mass term.

The lagrangian and hamiltonian for a massive scalar particle in this formalism are

$$\mathcal{L} = \partial\varphi_1/\partial x_\mu \partial\varphi_2/\partial x^\mu - m^2\ \varphi_1\varphi_2$$

with hamiltonian density

$$\mathcal{H} = \pi_1\ \pi_2\ + \partial\varphi_1/\partial x_i \partial\varphi_2/\partial x^i + m^2\ \varphi_1\varphi_2$$

The fields can be fourier expanded in terms of creation and annihilation operators:

$$\varphi_i(\mathbf{x}, t) = \int d^3k\ [a_i(k)f_k(x) +\ a_i^\dagger(k)f_k{}^*(x)]$$

for i = 1, 2 where

$$f_k(x) = e^{-ik\cdot x}/(2\omega_k(2\pi)^3)^{1/2}$$

with $\omega_k = |\mathbf{k}|$.

The creation and annihilation operators satisfy the commutation relations:

$$[a_i(k),\ a_j^\dagger(k')] = (1 - \delta_{ij})\delta^3(\mathbf{k} - \mathbf{k'})$$
$$[a_i(k),\ a_j(k')] = 0$$
$$[a_i^\dagger(k),\ a_j^\dagger(k')] = 0$$

for i, j = 1, 2.

In this formulation the defining properties of a physical state are:

$$\varphi_1(x)|\Phi, \Pi> = \Phi(x)|\Phi, \Pi>$$
$$\pi_1(x)|\Phi, \Pi> = \Pi(x)|\Phi, \Pi>$$

where $\Phi(x)$ and $\Pi(x)$ are sharp on the states and thus classical fields with

$$\Phi(\mathbf{x}, t) = \int d^3k\ [\alpha(k)f_k(x) +\ \alpha^*(k)f_k{}^*(x)]$$

and correspondingly for $\Pi(x)$.

4.2.7.3.3 Vacuum States for Scalar (Higgs) Particles with Non-Zero Vacuum Expectation Values

When we implement the mass mechanism, Φ is constant. We can define a set of states

$$a_1(k)|\alpha> = \alpha(k)|\alpha>$$
$$a_1^\dagger(k)|\alpha> = \alpha^*(k)|\alpha>$$

and correspondingly a set of coherent states

$$|\alpha> = C\exp\left\{\int d^3k\ [\alpha(k)a_2^\dagger(k) + \alpha^*(k)a_2(k)]\right\}|0>$$

where C is a normalization constant and where the vacuum state |0> satisfies

$$a_1(k)|0> = a_1^\dagger(k)|0> = 0$$

$$a_2(k)|0> \neq 0 \qquad\qquad a_2^\dagger(k)|0> \neq 0$$

The dual vacuum state satisfies

$$<0|a_2(k) = <0|a_2^\dagger(k) = 0$$
$$<0|a_1(k) \neq 0 \qquad\qquad <0|a_1^\dagger(k) \neq 0$$

With this coherent state formalism, which gives purely classical fields and yet also has quantum fields through the use of φ_2 and its creation and annihilation operators, we now have the machinery to define a mass mechanism without the introduction of a potential whose origin can only be described as dubious.

For we can define a coherent state for some k as

$$|\Phi, \Pi> = C\exp\{[(2\pi)^3\omega_k/2]^{1/2}\Phi[a_2^\dagger(k) + a_2(k)]\}|0>$$

where C is a normalization constant, that yields a non-zero vacuum expectation value:

$$\varphi_1(x)|\Phi, \Pi> = \Phi|\ \Phi, \Pi>$$

where Φ is a constant. Evaluating a fermion interaction term we find a mass term emerges[71]

$$\bar{\psi}(\varphi_1 + \varphi_2)\psi \ \rightarrow \bar{\psi}(\Phi + \varphi_2)\psi$$

It generates a mass for an interaction with a gauge field of the form

$$A^\mu(\varphi_1 + \varphi_2)^2A_\mu \ \rightarrow A^\mu(\Phi + \varphi_2)^2A_\mu$$

[71] When matrix elements with a "vacuum state" are taken.

It also yields a quantum field theoretic interaction that would result in the production of ElectroWeak particles from these scalar fields. The production of Higgs particles that decay into ElectroWeak gauge particles has recently been found at CERN.

The present formalism provides a clean way to separate the vacuum expectation value of a scalar particle from its quantum field part in contrast to the Higgs Mechanism where one has to separate a Higgs field into parts manually.

4.2.7.3.4 Interpretation of Negative Energy Scalar Particle States

As we noted earlier, scalar particle physics has the problem of no barrier to the decay of positive energy states to negative energy states due to the absence of a Pauli Exclusion Principle for bosons. The PseudoQuantization procedure that we developed in 1978 and describe here allows negative energy states as one would physically expect and raises the possibility of disastrous particle decays to negative energy. The above equations show that negative energy states are possible in this theory.

However they also show that combined positive and negative energy boson states can be interpreted as classical field states. In addition, the ability of any number of boson particles to have the same 4-momentum and quantum numbers shows that a *macroscopic* classical scalar field state can be constructed.

Thus we can view states containing negative energy particles as classical field states and thus solve[72] *the issue of interpreting negative energy particle states – a more satisfactory approach than the standard quantization procedure does – with due respect to Professor Dirac.*

We note that macroscopic many particle fermion states can only have one particle in any mode unlike bosons. Therefore we cannot use this formalism to create macroscopic classical fermion field states.[73] And the filled Dirac sea of negative energy fermions precludes the transition of a positive energy Dirac fermion to a negative energy state. *Thus there is a certain complementarity between fermions that cannot become classical fields but have a filled sea precluding decays to negative energy states, and bosons that can become classical fields but support decays to negative energy states.*

4.2.7.3.5 Contrast with Conventional Second Quantization of Scalar Particles

The PseudoQuantization procedure followed here uses different boundary conditions than the usual scalar particle quantization procedure. The essence of the difference is embodied in a comparison of the definition of the vacuum above and the definition of the conventional second quantized field vacuum:

---

[72] Also a boson that has no interactions cannot transition from to a positive energy state to a negative energy state due to conservation of energy.

[73] However we can create PseudoQuantum fermion states. See S. Blaha, Phys. Rev. **D17**, 994 (1978) and references therein to earlier papers by the author.

$$a|0> = 0 \qquad \text{Conventional Approach}$$
$$a^\dagger|0> \neq 0$$

In the conventional approach the creation of negative energy boson states is eliminated *ab initio* whereas in our approach it is allowed in order to support classical field states with non-zero vacuum expectation values that are a form of classical field. While one cannot discredit the conventional choice for conventional scalar fields, one can see that our approach yields a physically more important result – particularly for Higgs fields – because it leads to an Arrow of Time *locally* – an important feature of physical phenomena that has been a subject of much discussion and dispute. One can say that the conventional approach sweeps the issue "under the rug" rather than seeking a deeper justification – differing from Dirac's implied notion that the issue merited attention. We will discuss the "Arrow of Time" within the framework of our PseudoQuantization approach later.

4.2.7.3.6 WHY INERTIAL REFERENCE FRAMES ARE SPECIAL

The great physicists of the early 20th century raised numerous questions about Special Relativity after Einstein and Poincarè's discovery. Prominent among them was the question of why inertial reference frames are of especial importance in Special Relativity, and afterwards in General Relativity.

It appears that our formulation of the mass generation mechanism sheds significant light on the reason for the special prominence of inertial frames. Earlier we considered the case of a massless PseudoQuantized scalar. We now consider massive scalars since experiments at CERN have apparently discovered a Higgs particle with a 125 GeV/c mass. The above equations describe a massive scalar particle. If the scalar is massive, then the "vacuum" state that yields a non-zero expectation value must change to

$$|\Phi, \Pi> = C\exp\{(2\pi)^3 m/2]^{1/2}[a_2^\dagger(\mathbf{0},m) + a_2(\mathbf{0},m)]\}|0>$$

to have operators for a particle of mass m in its rest frame. Then, having established this preferred frame for a Higgs particle, in The Extended Standard Model, and requiring that invariant intervals

$$ds^2 = dt^2 - d\mathbf{x}^2 \quad \text{(in rectangular coordinates)}$$

are unchanged by a (complex or real) Lorentz transformation, we find that inertial reference frames are singled out as "special" in the sense that they are the only accessible reference frames that can be generated by a Lorentz boost/transformation from the Higgs particle rest frame. *The Higgs particle vacuum state singles out the class of inertial reference frames.*

Thus Higgs particles play a central role in establishing the basis of physical reality.

4.2.7.3.7 PSEUDOQUANTIZATION REVEALS MORE PHYSICAL CONSEQUENCES THAN THE HIGGS MECHANISM OF SCALAR PARTICLES

Earlier we pointed out that our PseudoQuantization theory of Higgs particles reveals more physical consequences than the conventional approach, which implements the Higgs Mechanism by simply using a potential term that has a minimum at a non-zero vacuum expectation value. This section shows the major results of a properly implemented mechanism. We find a better explanation of the negative energy state problem of boson field theories. We find a local arrow of time that explains the direction of time that we, and all of nature, experiences. We find the reason why inertial reference frames have a special physical significance – a result long sought by physicists.

In addition we will see in section 4.2.8 that real gauge fields should have an associated Higgs particle, while necessarily complex gauge fields (the Strong interaction gauge field in The Extended Standard Model) do not have an associated gauge field. These results correspond to experimental reality.

4.2.7.3.8 THE T INVARIANCE ISSUES OF OUR PSEUDOQUANTIZED SCALAR PARTICLE THEORY

The PseudoQuantized scalar particle hamiltonian equations are invariant under time reversal t → t' = –t. The 'new' vacuum states defined above break the time reversal invariance of the theory resulting in retarded particle propagators.

The hamiltonian equations

$$[H, \varphi_1(\mathbf{x}, t)] = -i\partial\varphi_1/\partial t$$
$$[H, \varphi_2(\mathbf{x}, t)] = -i\partial\varphi_2/\partial t$$

are invariant under time reversal. If we define a time reversal operator transformation U then the time reversed equations are

$$[UHU^{-1}, \varphi_1(\mathbf{x}, -t)] = +i\partial\varphi_1(\mathbf{x}, -t)/\partial(-t)$$
$$[UHU^{-1}, \varphi_2(\mathbf{x}, -t)] = +i\partial\varphi_2(\mathbf{x}, -t)/\partial(-t)$$

The operator U, which is unitary, transforms H into –H. This operation is legal because the hamiltonian – in this case – is not positive definite and admits negative energy states.[74] Thus

$$[H, \varphi_1(\mathbf{x}, -t)] = -i\partial\varphi_1(\mathbf{x}, -t)/\partial(-t)$$
$$[H, \varphi_2(\mathbf{x}, -t)] = -i\partial\varphi_2(\mathbf{x}, -t)/\partial(-t)$$

and the time reversal invariance of the equations of motion is established for this case.

Time reversal invariance is broken by our choice of vacuum states. This choice is necessary to obtain classical field states as we showed earlier. A demonstration of the time

[74] Unlike the usual case of second quantized Klein-Gordon quantum field theory.

reversal symmetry breaking is presented later where we show theory has retarded propagators for particle propagation to and from asymptotic states.

Within the interaction region the particle propagators are the sum of retarded and advanced parts that combine to yield principle value propagators – not Feynman propagators. Many years ago Feynman and Wheeler championed principle value propagators for electrodynamics to obtain an action-at-a distance theory of Quantum Electrodynamics. While their theory, and ours, differ from the standard quantum field theory approach there is no reason to view them as faulty, or having serious physical defects. The only question is whether nature chooses conventional quantum field theory or PseudoQuantized quantum field theory. In our case the need for a classical scalar particle non-zero vacuum expectation value strongly motivates our choice of psedoquantized Higgs particles.

4.2.7.3.9 Retarded Propagators for Our Quantized Higgs Particles

In the previous section we pointed out that our PseudoQuantization Higgs theory has an arrow of time due to is boundary conditions as expressed by its definition of the vacuum state and its dual. In this section we will show that the theory uses retarded propagators for propagation to and from the interaction region to asymptotic in-states and out-states. Within an interaction region the theory uses half-retarded – half-advanced propagators. We discuss aspects of the perturbation theory and propagators of our scalar particles in this chapter.

First we note that in-states at $t = -\infty$ are composed of superpositions of $a_2(k)$ and $a_2^\dagger(k)$ creation and annihilation operators:

$$a_2(k)|0> \neq 0 \qquad\qquad a_2^\dagger(k)|0> \neq 0$$

while the out-states composed of superpositions of $a_1(k)$ and $a_1^\dagger(k)$ creation and annihilation operators:

$$<0|a_1(k) \neq 0 \qquad\qquad <0|a_1^\dagger(k) \neq 0$$

Consequently when in-state particles (x_1) propagate into the interaction region (x_2) the relevant propagators are retarded propagators with the form

$$\begin{aligned} G_{in}(x_2, x_1) &= <0|T(\varphi_{1\,in}(x_2), \varphi_{2\,in}(x_1))|0> \\ &= \theta(x_{20} - x_{10})<0|[\varphi_{1\,in}(x_2), \varphi_{2\,in}(x_1)]\,|0> \end{aligned}$$

This is a manifestly retarded propagator. The choice of vacuums clearly results in a time asymmetry giving a retarded propagation reflecting the familiar Arrow of Time.

A similar situation prevails for propagation to out-states (x_3) from the interaction (x_2) region:

$$\begin{aligned} G_{out}(x_3, x_2) &= <0|T(\varphi_{1\,out}(x_3), \varphi_{2\,out}(x_2))|0> \\ &= \theta(x_{30} - x_{20})<0|[\varphi_{1\,out}(x_3), \varphi_{2\,out}(x_2)]\,|0> \end{aligned}$$

Within the interaction region the Higgs particles have principle value propagators.

Thus we find PseudoQuantized Higgs particles embody a local Arrow of Time. The locality of the Arrow of Time is embodied in all the particles that interact with the Higgs particle. Since the mass of *every* particle – bosons and fermions – has a Higgs contribution, and thus *every* particle interacts with the Higgs particles, the Arrow of Time permeates The Unified SuperStandard Model as well as the more familiar Standard Model known from experiment.

4.2.7.3.10 THE LOCAL ARROW OF TIME

In the *Physics is Logic* series of monographs we saw that complex coordinates led to the form of the fermion spectrum, that the mapping of complex coordinates to real-valued coordinates yielded the Reality group and The Extended Standard Model gauge interactions, that Complex General Relativity led to Higgs particles that were directly united with elementary particle masses and gave us the equality of inertial mass and gravitational mass. In section 4.2.8 we wil see the reduction of complex gauge fields to real gauge fields explains the appearance of Higgs fields in The Unified SuperStandard Model.

The PseudoQuantization procedure leads to retarded Higgs field propagators and thence to a *local* arrow of time. Many arguments have been put forward over the past hundred plus years for the Arrow of Time. Many arguments based on Statistical Mechanics, Entropy, and Boltzmann's statistical atomic theory have suggested the Arrow of Time is a global statistical consequence. This view seems to contradict the results of elementary particle experiments where a *local* Arrow of Time is evident.

Our rationale for the Arrow of Time begins with retarded Higgs fields. Then we note that Higgs field quantum interactions appear for all fermions and gauge particles. Thus all particle interactions are imbued with an Arrow of Time. Particles united to form macroscopic matter inherit their combined Arrows of Time producing the global Arrow of Time we experience.

Thus our PseudoQuantization approach offers a more satisfactory solution of the origin of the Arrow of Time.

It is remarkable that complex quantities – coordinates and fields – through the Higgs phenomena that we have considered, lead to the equality of inertial mass and gravitational mass, and an Arrow of Time. This unity of mass and time phenomena may reflect the deeper fact that we can have no practical Arrow of Time if all particles were massless, for particle dynamics at light speed would then be pointless. This view has been expressed by DeWitt, Unruh, and others who have pointed out that, physically, time is meaningful and measurable only if masses exist; the larger the mass, the more accurate the time measurement in principle.[75]

---

[75] No mass, no clock; no clock, no physical time. See Blaha (2015a) pp. 368-371 for a discussion including comments by DeWitt and Unruh.

4.2.7.3.11 Space-Time Dependent Particle Masses

It is possible that the ultimate Unified SuperStandard Model has masses that evolve with time and may also be spatially varying – different values in different parts of the universe. Presently there is no decisive evidence for this possibility although astrophysical studies continue. In this section we will describe the mechanism for space-time dependent masses.

Consider a classical field (time and spatially varying):

$$\Phi(\mathbf{x}, t) = \int d^3k\, [\alpha(k)f_k(x) + \alpha^*(k)f_k{}^*(x)]$$

If we define the coherent vacuum state

$$|\alpha\rangle = C \exp\left\{\int d^3k\, [\alpha(k)a_2{}^\dagger(k) + \alpha^*(k)a_2(k)]\right\}|0\rangle$$

then

$$\varphi_1(x)|\Phi, \Pi\rangle = \Phi(x)|\Phi, \Pi\rangle$$
$$\pi_1(x)|\Phi, \Pi\rangle = \Pi(x)|\Phi, \Pi\rangle$$

where

$$\varphi_i(\mathbf{x}, t) = \int d^3k\, [a_i(k)f_k(x) + a_i{}^\dagger(k)f_k{}^*(x)]$$

for i = 1, 2 and where

$$f_k(x) = e^{-ik\cdot x}/(2\omega_k(2\pi)^3)^{1/2}$$

with ω_k equal to the energy.

4.2.7.3.12 Inertial Mass Equals Gravitational Mass

From the days of Newton through Einstein[76] to the present the equality of gravitational mass and inertial mass has been a topic of interest. Mach, who played an important role, in this ongoing discussion, thought distant masses in the universe were the source of the equality. However the origin of the equality, which has been shown experimentally to very high accuracy, remained uncertain until the *Physics is Logic* series of books, in which we showed the interconnection of the Extended Standard Model and Complex Gravitation via Higgs generated masses that united gravitational and inertial mass.

In Blaha (2016h) we showed that a Complex General Relativity transformation can be factored into the product of a complex-valued transformation and a real-valued General Coordinate transformation. The set of complex valued transformations form a U(4) group that we called the General Coordinate Reality group. This group has gauge fields that undergo spomtaneous symmetry breaking and generate contributions to all fermion masses.

[76] For example, Einstein and Grossman in 1913 stated, "The theory herein described originates in the conviction that the proportionality between the inertial and gravitational mass of a body is an exact law of nature that must be expressed as a foundation principle of theoretical physics."

Since fermion field masses are now sums of ElectroWeak Higgs contributions, Generation group Higgs contributions, Layer group Higgs contributions, and General Coordinate Species group contributions, and since the gravitational Higgs fields appear in all fermion masses, the equality of inertial and gravitational mass is proven. The gravitational Higgs particles' equations depend, in part, on the gravitational field by Blaha (2016h) and so set the mass scale of gravitational mass, and thereby of all Higgs mass contributions. They set the scale of inertial masses equal to the scale of gravitational masses. ***Since an expression cannot mix mass scales, the gravitational mass scale must be the same as the inertial mass scale. Inertial Mass equals gravitational mass.***

We have established the equality of inertial and gravitational mass at the short distance quantum level. In our view, this explanation is far more satisfying than basing the equality on a combination of large distance phenomena and quantum phenomena. As Einstein and Weyl have pointed out, all fundamental physics phenomena should be based on a local theory. Complex Gravity as we have constructed it, combined with the Extended Standard Model, furnishes a completely local basic Theory of Everything.

The equation above contains a coherent state $|\alpha>$ for a time and spatially varying mass. The above equations can be generalized to the case of multiple space-time varying masses.[77]

$$|\Phi_1,\Phi_2, \ldots ,\Phi_n;\Pi_1,\Pi_2, \ldots ,\Pi_n> = C \prod_{i=1}^{n} \exp\left\{\int d^3k\, [\alpha_i(k)a_{2i}^{\dagger}(k) + \alpha_i^{*}(k)a_{2i}(k)]\right\}|0>$$

Then all n mass vacuum expectation values are space-time dependent:

$$\varphi_{1i}(x)\,|\,\Phi_1, \Phi_2, \ldots , \Phi_n; \Pi_1, \Pi_2, \ldots , \Pi_n> = \Phi_i(x)\,|\,\Phi_1, \Phi_2, \ldots , \Phi_n; \Pi_1, \Pi_2, \ldots , \Pi_n>$$

Thus our formalism can accommodate space-time varying masses should they be found in the Cosmos.

4.2.7.3.13 BENEFITS OF THE PSEUDOQUANTIZATION METHOD

In this book and earlier work we showed that a more physically satisfactory method for avoiding the negative energy state problem exists. This method relies on the use of a larger Fock space in which negative energy states (or partially negative energy states) are interpreted as states containing classical fields or a mix of classical fields and individual boson particles. This approach resolves the negative energy boson issue and provides a common framework for boson particles and classical boson fields.

One consequence of the PseudoQuantization method is that it enables the appearance of a vacuum expectation value for Higgs particles (a constant classical field) to be understood within a truly quantum framework. Another major consequence of this approach is the

[77] The "vacuum" state $|0>$ also implicitly has factors for the vacuum expectation values used for fields that give masses to fermions and vector bosons as described in Blaha (2016h).

appearance of a *local* Arrow of Time due to the Higgs mass generation mechanism – a concept that has been a subject of interest for over one hundred years. A macroscopic arrow of time is often described as a statistical result. But our approach yields an arrow of time at the single particle level.

The conventional approach to boson field quantization sweeps these issues "under the rug" rather than seeking a deeper justification. It differs from Dirac's implied notion that the issue merited attention.

Another important consequence of the PseudoQuantization method is that it singles out inertial reference frames when applied to the case of Higgs particles.

Yet another more subtle consequence of boson PseudoQuantization is that it provides a rationale/explanation for the presence of ElectroWeak Higgs bosons, *and for their absence for the strong (gluon) interactions. The question of why there are no strong interaction Higgs bosons has not been previously considered to the best of this author's knowledge.*

4.2.8 Higgs Particles, Gauge Fields, and Higgs Mechanism Generated Coupling Constants

Higgs particles appear in many contexts in the Unified SuperStandard Model. In this subsection we consider a possible origin of Higgs particles in complex-valued gauge fields that explains why there are no Strong Interaction gauge field Higgs particles. We also show it is possible to define Higgs particles that generate gauge field coupling constants. Using this mechanism we show that the known coupling constants appear to have values of the same order of magnitude suggesting that we are close to a form of unification. We also show the Higgs Mechanism for fermion particle masses may explain the equality of inertial and gravitational mass – a topic of continuing interest for many years.

4.2.8.1 The Genesis of Scalar (Higgs?) Particle Fields from Complex Gauge Fields

Sections 4.2.8.1- 4.2.8.3 show that scalar particles can be 'extracted' from all spin 1 gauge fields except color SU(3).[78]

Since our Unified SuperStandard Model is ultimately based on the Complex General Coordinate Transformations and the Complex Lorentz group (thus complex-valued coordinate systems), it appears reasonable to consider all spin 1 gauge fields to initially be similarly complex-valued. Most of these gauge fields can be rotated to real values. However we shall see that color SU(3) gauge fields are *necessarily* complex-valued. All other gauge fields can be rotated to real values. The price of rotation is the introduction of scalar fields. Some of these fields may be Higgs particle fields and generate gauge boson masses (symmetry breaking) and fermion masses.

Thus we can view scalar particles including Higgs particles as inherently associated with most gauge fields. From the viewpoint of our derivation from basic axioms, the origin of Higgs bosons in complex-valued gauge fields gives a 'tighter' derivation of the overall theory. Higgs fields are inherently a natural part of the theory.

4.2.8.2 The Difference between the Strong Gauge Field and the Other Gauge Fields in the SuperStandard Model

In our Unified SuperStandard Model the only gauge field without an associated Higgs particle is the strong interaction gluon gauge field. *We view this exception as a particularly important clue as to the nature of the relation between gauge fields and Higgs particles.*

How does the strong interaction gauge field differ from all other gauge fields in the Unified SuperStandard Model? An examination of the gauge fields dynamic equations (and other lagrangian terms) of our Unified SuperStandard Model reveals that all gauge field dynamic equation kinetic terms *except the strong interaction gauge field* have the form:

---

[78] The concepts of sections 4.2.8.1 – 4.2.8.3 first appeared in Blaha (2015c) and (2016c).

$$\partial/\partial x_\mu F^a{}_{\mu\nu} + gf^{abc}A^{b\mu} F^c{}_{\mu\nu} = j^a{}_\nu$$

where

$$F^a{}_{\mu\nu} = \partial/\partial x^\nu A^a{}_\mu - \partial/\partial x^\mu A^a{}_\nu + gf^{abc}A^b{}_\mu A^c{}_\nu$$

where the coordinates x^ν *are real-valued,*[79] where a, b, c are structure constant indices, where g is a coupling constant, and where $j^a{}_\nu$ is the corresponding current. The gauge field $A^a{}_\mu$ is real for all ElectroWeak gauge fields, Generation group gauge fields, and Layer group gauge fields. Thus the above equations are real-valued.

The strong interaction gauge field[80] in our Unified SuperStandard Model differs from the other gauge fields by being *necessarily* complex[81] due to the complex 3-space complexon derivatives that appear in the corresponding dynamic equations:

$$D^\mu F_C{}^a{}_{\mu\nu} + gf^{abc}A_C{}^{b\mu} F_C{}^c{}_{\mu\nu} = j^a{}_\nu$$

with

$$F_C{}^a{}_{\mu\nu} = D_\nu A_C{}^a{}_\mu - D_\mu A_C{}^a{}_\nu + gf^{abc}A_C{}^b{}_\mu A_C{}^c{}_\nu$$

where

$$D_k = \partial/\partial x_r{}^k + i\, \partial/\partial x_i{}^k$$
$$D_0 = \partial/\partial x^0$$

for k = 1, 2, 3 where $A_C{}^a{}_\mu$ is the complexon color Strong interaction gauge field. The complexon spatial coordinates have the form $x_r{}^k + i\, x_i$. The time coordinate is real-valued. These equations are eqs. 10.16 and 5.162 of Blaha (2015a) for complexon gauge fields,[82] which carry the strong interaction in the Unified SuperStandard Model.

This difference enables us to differentiate the strong gauge field from all other gauge fields in The Unified SuperStandard Model. Thereby we can develop a unified formalism for the non-strong gauge fields and their corresponding Higgs particles.

The necessarily complex nature of the color SU(3) field is the reason that the Strong Interaction gauge fields do not acquire a mass via the Higgs Mechanism. As shown below, the necessary complexity of Strong Interaction gauge fields precludes the generation of Higgs fields from Yang-Mills gauge fields.

---

[79] Before the introduction of the Two-Tier formalism.
[80] This field is called a complexon gauge field in Blaha (2017b), (2015a) and earlier books.
[81] One cannot cleanly separate the real and imaginary parts of its dynamic equations.
[82] In The Unified SuperStandard Model we also identify quark species particles as having complex 3-momentum. We call them complexon fermions.

4.2.8.3 Generation of Higgs Fields from Non-Abelian Gauge Fields

In the prior section we considered the difference between the strong gauge field and the other gauge fields of The Unified SuperStandard Model. Unlike strong gauge fields the other gauge fields (ElectroWeak and so on) could be real or complex. In a manner similar to what we did in the preceding *Physics is Logic* books (and earlier books) we can assume gauge fields are initially complex, and then transform them to real-valued fields using a phase transformation that introduces scalar fields, some of which we will take to be Higgs fields.

We define a complex phase transformation for a gauge field $A^{b\mu}$ with

$$A'^{a\mu}(x) = \Phi(x)^{a}{}_{b}A^{b\mu}(x)$$

where $\Phi(x) = \text{diag}(\exp[i\varphi_1(x)], \exp[i\varphi_2(x)], \ldots, \exp[i\varphi_n(x)])$, and n is the number of symmetry components of $A^{b\mu}$. Inserting $A'^{a\mu}(x)$ above we find:

$$\partial/\partial x_\mu F'^{a}{}_{\mu\nu} + gf^{abc}A'^{b\mu} F'^{c}{}_{\mu\nu} = j^{a}{}_{\nu}$$

where

$$F'^{a}{}_{\mu\nu} = \partial/\partial x^{\nu}\{\exp[i\varphi_a(x)]A^{a}{}_{\mu}\} - \partial/\partial x^{\mu}\{\exp[i\varphi_a(x)]A^{a}{}_{\nu}\} + gf^{abc}\exp[i\varphi_b(x)]\, A^{b}{}_{\mu}\exp[i\varphi_c(x)]A^{c}{}_{\nu}$$

If we now assume that $\varphi_a(x)$ is small for all a then

$$\exp[i\varphi_a(x)] \simeq 1 + i\varphi_a(x)$$

to first order. Substituting above, and keeping terms to leading order yields the real part:

$$\partial/\partial x_\mu F^{a}{}_{\mu\nu} + gf^{abc}A^{b\mu} F^{c}{}_{\mu\nu} = j^{a}{}_{\nu}$$

where $F^{a}{}_{\mu\nu}$ is given above, and the imaginary part is:

$$\partial/\partial x_\mu F_{i}{}^{a}{}_{\mu\nu} + gf^{abc}A^{b\mu} F_{i}{}^{c}{}_{\mu\nu} = 0$$

to leading order where

$$F_{i}{}^{a}{}_{\mu\nu} = \partial/\partial x^{\nu}\, \varphi_a(x)A^{a}{}_{\mu} - \partial/\partial x^{\mu}\, \varphi_a(x)A^{a}{}_{\nu}$$

Then we find

$$A^a_\nu \Box \varphi_a(x) - A^a_\mu\, \partial/\partial x_\mu \partial/\partial x^\nu\, \varphi_a(x) - g f^{abc} A^{b\mu}\, [A^c_\mu\, \partial/\partial x^\nu\, \varphi_a(x) - A^c_\nu\, \partial/\partial x^\mu\, \varphi_a(x)] = 0$$

in the Lorentz gauge, with no sum over a. This equation is a form of Klein-Gordon equation having interaction terms with the gauge field. If the gauge field is weak then only the first two terms are important.

Note that only derivatives of $\varphi_a(x)$ appear above. Consequently shifts of the $\varphi_a(x)$ field by a constant still yield solutions. This feature makes $\varphi_a(x)$ a candidate to be a Higgs particle field.

Note also that complexon gauge fields cannot have such a phase change, with a subdivision into real and imaginary dynamic equations, due to the complexity of the spatial coordinates. This difference appears to be the reason why the strong interaction gauge field does not have an associated Higgs particle.

The $\varphi_a(x)$ particles can be made into Higgs particles by adding an appropriate potential:

$$V = A\, \varphi_a^2(x) + B\, \varphi_a^4(x)$$

where A and B are constants. Approximating with its first two terms and inserting the potential term we find the Higgs-like equation:

$$A^a_\nu \Box \varphi_a(x) - A^a_\mu\, \partial/\partial x_\mu \partial/\partial x^\nu\, \varphi_a(x) + \partial V/\partial \varphi_a = 0$$

$\varphi_a(x)$ has a minimum at the minimum of the potential in the corresponding lagrangian. The second and third terms constitute the interaction. Neglecting these terms we see that we obtain the free, massless, field Klein-Gordon equation

$$\Box \varphi_a(x) = 0$$

The pairing of Higgs particles with real-valued gauge fields is thus established.[83] The non-existence of a matching Higgs field for the strong interaction is due to the inherently complex nature of the strong interaction (complexon) gauge field in the Unified SuperStandard Model also follows.

The derivation presented here is analogous to the derivation of Higgs fields in Complex General Relativity – also a gauge theory – in *Physics is Logic Part II.*

One of the remarkable aspects of The Unified SuperStandard Model is its ability to directly prove qualitative properties of elementary particles: four fermion species, Parity

[83] Some of the Higgs fields so generated may not have vacuum expectation values and so may only play a role in interactions.

violation, the distinction between leptons and quarks, the match of the SuperStandard Model's (broken) symmetries with the internal symmetry Reality group, and now the existence of Higgs gauge fields in all interaction sectors except for the strong interactions. We take these successes to be indicators of the correctness of The Unified SuperStandard Model.

4.2.8.4 General Higgs Formulation of Gauge and Fermion Particle Masses

We have seen seven of the interactions present in our Unified SuperStandard Model. Four more interactions will be presented later. One of them is the Species gauge field $A_S{}^{\mu}$ generated from the Reality group of complex General Relativistic transformations. There are two more interactions associated with General Relativistic transformations and an all-encompassing interaction $A_\Theta{}^{\mu}(x)$ considered later. We shall discuss the Higgs Mechanism associated with seven of the spin 1 gauge field interactions that appear in fermion covariant derivatives. They can be put in a vector form:[84]

$$\mathbf{A}_I{}^{\mu} = (g_1\mathbf{A}_{SU(3)}{}^{\mu}(x_C), g_2\mathbf{W}^{\mu}(x)\,, g_3\mathbf{A}_E{}^{\mu}(x), g_4\mathbf{W}_D{}^{\mu}(x), g_5\mathbf{A}_{DE}{}^{\mu}(x), g_6\mathbf{U}^{\mu}(x), g_7\mathbf{V}^{\mu}(x))$$

where each element is a vector of the gauge fields in the group of the gauge field and the respective coupling constants are labeled $g_1, g_2, \ldots, g_7$. The subscript 'D' labels Dark matter interactions. 'W' labels Weak fields, 'E' labels Electromagnetic fields. 'U' labels U(4) Generation group fields., and 'V' labels U(4) Layer group fields.

The interactions' symmetry is $[SU(3)\otimes SU(2)\otimes U(1)\otimes SU(2)\otimes U(1)\otimes U(4)\otimes U(4)]^4$ since, as we shall see, each of the four fermion layers due to the Layer groups discussed later, has a separate set of the seven interactions.[85] In each layer the number of fields for the seven interactions is 8. 3, 1, 3, 1, 16, and 16 – totaling 48 fields.

Similarly we define a 7-vector of 48 generators

$$\mathbf{T}_I = (\mathbf{T}_{SU(3)}, \boldsymbol{\tau}_{SU(2)}, \mathbf{I}_{U(1)}, \boldsymbol{\tau}_{DSU(2)}, \mathbf{I}_{DU(1)}, \mathbf{G}_{U(4)}, G_{LU(4)})$$

Then the total gauge fields interaction term within a covariant derivative corresponding to the seven interactions, the General Relativistic U(4) Species interaction, the spinor interaction, and the Θ-interaction can be expressed as

$$\mathbf{A}_I{}^{\mu}{}_k\mathbf{T}_{Ik} + A_S{}^{\mu}{}_k\mathbf{G}_{Sk} + g_B B^{\mu} + g_\Theta A_\Theta{}^{\mu}(x)$$

summed separately over k for each interaction. The remaining additional interactions are real-valued gravitational connections that we will describe later. The covariant derivative of a fermion field (neglecting General Relativistic terms) for each layer is

[84] Later we will reformulate this discussion in terms of PseudoQuantum field theory.

[85] Otherwise the various layers would have interactions between them which would have appeared in experiments. Then the upper three layers would not be Dark.

$$\{\partial^\mu + i\,[\mathbf{A}_I{}^\mu + A_S{}^\mu + g_B B^\mu + g_\Theta A_\Theta{}^\mu(x)]\}\gamma_\mu\psi = 0$$

Note the complexon nature of the SU(3) gauge field makes us use the covariant derivative

$$\{\partial/\partial x_{C\mu} + i\,[\mathbf{A}_I{}^\mu + A_S{}^\mu + g_B B^\mu + g_\Theta A_\Theta{}^\mu(x)]\}\gamma^\mu\psi = 0$$

for the SU(3) quark dynamic equations where the other gauge fields are functions of $x_r = \text{Re } x_C$.

We now consider the combined effects of the seven interactions, $\mathbf{A}_I{}^\mu$, on generating gauge boson masses (symmetry breaking) and fermion masses.[86] We begin by defining a composite Higgs field for all 7 interactions:

$$\eta = \prod_{k=1}^{7} \eta_{kTSLg}$$

where k labels the group, T labels the type of matter, S labels the species, L labels the layer and g labels the generation. We now consider

$$D^\mu\eta = \{\partial^\mu + i\,[\mathbf{A}_I{}^\mu + g_B B^\mu + g_\Theta A_\Theta{}^\mu(x)]\}\eta$$

Letting η be a product of real fields ρ.., whose elements are composed of zeroes and non-zero real fields

$$\eta = \prod_{k=1}^{7} \rho_{kTSLg}$$

Then we find that

$$(D_\mu\eta)^\dagger D^\mu\eta = \sum_{kTSLg} \partial_\mu\rho_{kTSLg}\,\partial^\mu\rho_{kTSLg} + \sum_{kTSL} g_k{}^2\beta_{kTSL}U_{kTSL}{}^2$$

where β_{kTSL} is a sum of terms, each of which is quadratic in the vacuum expectation value of a Higgs field. The second term above yields the masses of the gauge fields of non-zero mass.

The lagrangian terms that generate fermion masses have the form

$$\mathscr{L}_{\text{FermionMasses}} = \sum_{kTSLgh} \bar{\psi}_{L\,kTSLg}\,\rho_{0kTSL}\,m_{kTSLgh}\,\psi_{R\,kTSLh}$$

---

86 *We develop the Higgs Mechanism in detail in Blaha (2017b) and earlier books for fermions and bosons for all seven interactions above plus the Reality group $A_S{}^\mu$ of Complex General Relativistic transformations, the Generation group, the Layer groups, and the Θ-group gauge field $A_\Theta{}^\mu(x)$.*

where the sums over g and h are over the generations of a specific T, S, and L of a group labeled k, and ρ_{0kTSL} is the vacuum expectation value of a Higgs particle. The initial 'L' and 'R' subscripts represent Left and Right.

We note that the fermion mass matrices can be diagonalized using a matrix A_{TSL}:

$$m_{TSLphys} = A_{L\,TSL} \sum_k \rho_{0kTSL} m_{kTSL} A_{R\,TSL}^{-1}$$

where $m_{TSLphys}$ is the diagonal mass matrix for the generations specified by T, S, and L.

Thus the lagrangian fermion mass terms for physical fermions become

$$\mathscr{L}_{FermionMasses} = \sum_{TSL} \bar{\psi}_{LphysTSL}\, m_{TSLphys} \psi_{RphysTSL} + c.c.$$

with diagonal mass matrices $m_{TSLphys}$.

4.2.8.5 The Mixing Pattern in the Fermion Periodic Table

The preceding discussions describe the pattern of mixing resulting from the ElectroWeak, Generation group, and Layer group. Fig. 4.2 pictorially presents an example of the mixing pattern within the Periodic Table of Fermions.

4.2.8.6 The Full Unified SuperStandard Model Fermion Mass Matrices

The above equations lead to the total mass matrices *for the four layers* listed below. The masses for each particle in the various layers are different although we use the same symbol for each type of mass. We include the mass contributions, m_{Gi}, from the Species group with gauge fields $A_S{}^{\mu}$.

The below list is for one layer. The other three layers have a similar pattern. The masses listed in the list are symbolic and are not of the same value for each particle type. We denote them generally as: m_{Wi} for the Weak group contribution, m_{Li} for the Layer group contribution, m_{Geni} for the Generation group contribution, and m_{Gi} for the (Gravitational) Species group contribution with gauge fields $\mathbf{A}_S{}^{\mu}$.

Charged Lepton Species Total Mass Matrix

$$m_{etot} = m_{We} + m_{Le} + m_{Ge}$$

Neutral Lepton Species Mass Matrix

$$m_{\upsilon tot} = m_{W\upsilon} + m_{L\nu} + m_{G\upsilon}$$

Up-Type Quark Species Mass Matrix (for each color)

$$m_{utot} = m_{Wu} + m_{Lu} + m_{Genu} + m_{Gu}$$

Down-Type Quark Species Mass Matrix (for each color)

$$m_{dtot} = m_{Wd} + m_{Ld} + m_{Gend} + m_{Gd}$$

Dark Charged Lepton Species Total Mass Matrix

$$m_{Detot} = m_{DWe} + m_{DLe} + m_{Ge}$$

Dark Neutral Lepton Species Mass Matrix

$$m_{D\upsilon tot} = m_{DW\upsilon} + m_{DL\upsilon} + m_{G\upsilon}$$

Dark Up-Type Quark Species Mass Matrix

$$m_{Dutot} = m_{DWu} + m_{DLu} + m_{DGenu} + m_{Gu}$$

Dark Down-Type Quark Species Mass Matrix

$$m_{Ddtot} = m_{DWd} + m_{DLd} + m_{DGend} + m_{Gd}$$

THE FERMION PERIODIC TABLE

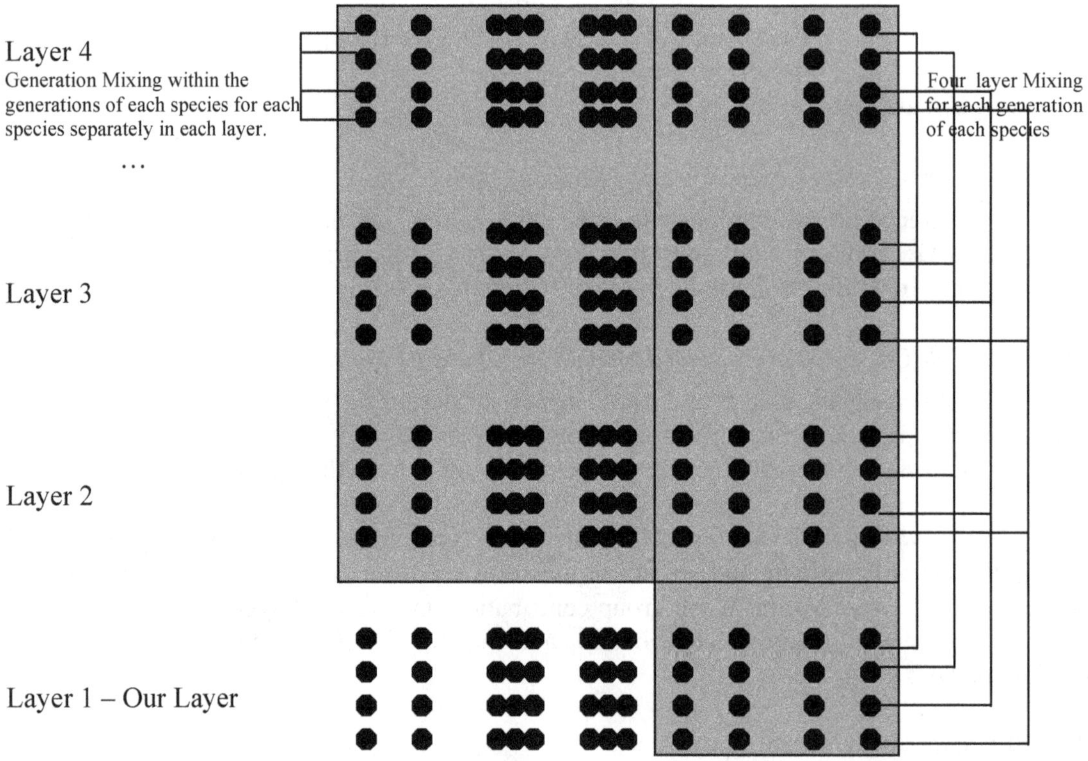

Figure 4.2. Partial example of pattern of mass mixing of the Generation group and of the Layer group. Dark parts of the periodic table are gray. Light parts are the known fermions with an additional, as yet not found, 4th generation shown. The lines on the left side show an example of the Generation mixing within one species. The Generation mixing applies to each species in each layer. The lines on the right side show an example of Layer mixing within one species with the mixing amongst all four layers of the species for each generation individually.

The (gravitational) Species group contribution to each fermion mass, m_{Gi} for each fermion type i, sets the scale for all fermion masses (and secondarily of massive gauge bosons' masses) yielding the "principle" of Newton, Einstein and others that *inertial mass equals gravitational mass.* *NOTE: The generation group contributions, in the spontaneous breakdown that we described, appear only in quark and Dark quark mass matrices possibly providing a reason why quark masses are so much larger than lepton masses. See Blaha (2017b).*

4.2.8.7 Higgs Mechanism for Coupling Constants

Particle masses are attributed to the operation of the Higgs Mechanism. Coupling constants are also constants that appear in the Unified SuperStandard Model. In this section[87] we define a Higgs Mechanism that yields the values of coupling constants as vacuum expectation values of Higgs particles.

4.2.8.7.1 THE INTERACTION COUPLING CONSTANTS AND THEIR PSEUDOQUANTUM FIELD VACUUM EXPECTATION VALUES

Ten of our Unified SuperStandard Model coupling constants are:[88]

- The Strong interaction coupling constant field g_S.
- The Weak SU(2) coupling constant g_W.
- The Electromagnetic U(1) coupling constant g_E.
- The Dark Weak SU(2) coupling constant g_{DW}.
- The Dark Electromagnetic U(1) coupling constant g_{DE}.
- The U(4) Generation group coupling constant g_G.
- The U(4) Layer group coupling constant g_V.
- The U(4) Species group coupling constant g_S.
- The U(192) Θ-interaction group coupling constant g_Θ.
- The complex gravitational coupling constant $g_{GR} = \kappa^{-1} = (4\pi G)^{-1/2}$.

Based on the discussions of Blaha (2015d) we can define Higgs vacuum expectation values for these coupling constants using a mass factor to obtain the correct coupling constant

---

[87] This chapter is largely extracted from Blaha (2015d).

[88] The following groups and their coupling constants are duplicated four-fold for the four layers. The Strong interaction coupling constant field g_S, The Weak SU(2) coupling constant g_W, The Electromagnetic U(1) coupling constant g_E, The Dark Weak SU(2) coupling constant g_{DW}, The Dark Electromagnetic U(1) coupling constant g_{DE}, The U(4) Generation group coupling constant g_G, and The U(4) Layer group coupling constant g_V. We will only consider the known first layer here The following group constants appear once – the same for all four layers – the U(4) Species group, Gravitational coupling constant, and the Θ-symmetry group. The interactions of these groups will be discussed later.

dimensions. We use the PseudoQuantum formalism of section 4.2.7.3.3 where we set the constants as follows:

- The Strong interaction coupling constant field $\Phi_1 = m_1 g_S$.
- The Weak SU(2) coupling constant $\Phi_2 = m_2 g_W$.
- The Electromagnetic U(1) coupling constant $\Phi_3 = m_3 g_E$.
- The Dark Weak SU(2) coupling constant $\Phi_4 = m_4 g_{DW}$.
- The Dark Electromagnetic U(1) coupling constant $\Phi_5 = m_5 g_{DE}$.
- The U(4) Generation group coupling constant $\Phi_6 = m_6 g_G$.
- The U(4) Layer group coupling constant $\Phi_7 = m_7 g_V$.
- The U(4) Species group coupling constant $\Phi_8 = m_8 g_S$.
- The U(192) Θ-interaction group coupling constant $\Phi_9 = m_9 g_\Theta$.
- The complex gravitational coupling constant $\Phi_{10} = m_{10} g_{GR} = \kappa^{-1} = (4\pi G)^{-1/2}$.

The ten masses, m_1 , m_2 , … , m_{10} may be equal or they may have different values. It is also possible they all may be equal to κ^{-1}, which would yield

- The Strong interaction coupling constant field $\Phi_1 = \kappa^{-1} g_S$.
- The Weak SU(2) coupling constant $\Phi_2 = \kappa^{-1} g_W$.
- The Electromagnetic U(1) coupling constant $\Phi_3 = \kappa^{-1} g_E$.
- The Dark Weak SU(2) coupling constant $\Phi_4 = \kappa^{-1} g_{DW}$.
- The Dark Electromagnetic U(1) coupling constant $\Phi_5 = \kappa^{-1} g_{DE}$.
- The U(4) Generation group coupling constant $\Phi_6 = \kappa^{-1} g_G$.
- The U(4) Layer group coupling constant $\Phi_7 = \kappa^{-1} g_V$.
- The U(4) Species group coupling constant $\Phi_8 = \kappa^{-1} g_S$.
- The U(192) Θ-interaction group coupling constant $\Phi_9 = \kappa^{-1} g_\Theta$.
- The complex gravitational coupling constant $\Phi_{10} = \kappa^{-1} g_{GR} = \kappa^{-1} = (4\pi G)^{-1/2}$.

Then scaling the above vacuum expectation values by κ^{-1} would give:[89]

- The strong interaction coupling constant[90] vacuum expectation value $\Phi_1' = g_S = 1.22$
- The Weak SU(2) coupling constant vacuum expectation value $\Phi_2' = g_W = 0.619$.
- The Electromagnetic U(1) coupling constant vacuum expectation value $\Phi_3' = e = g_E = 0.303$.
- The Dark Weak SU(2) coupling constant vacuum expectation value $\Phi_4' = \Phi_2'$. (?)
- The Dark Electromagnetic U(1) coupling constant vacuum expectation value $\Phi_5' = = \Phi_3'$. (?).
- The U(4) Generation group coupling constant $\Phi_6' = g_G$.
- The U(4) Layer group coupling constant $\Phi_7' = g_V$.
- The U(4) Species group coupling constant $\Phi_8' = g_S$.

---

[89] All coupling constant values are based on data extracted from K. A. Olive et al (Particle Data Group), Chinese Physics **C38**, 090001 (2014).
[90] Based on the running coupling constant value $\alpha_s(M_Z^2) = 0.1193 \pm 0.0016$.

- The U(192) Θ-interaction group coupling constant $\Phi_9' = g_\Theta$.
- The complex gravitational coupling constant $\Phi_{10}' = 1$.

The known *scaled* vacuum expectation values,[91] which are in fact the coupling constants, have a comparable range of values[92] as opposed to the range of values for the unscaled constants which range from the ultra-small gravitational vacuum expectation value to values, perhaps, within a few orders of magnitude of unity.

Given the range of known values above, it appears reasonable to conjecture that the unknown values would also be of the order of unity.

The known coupling constant values above are of comparable value, which suggests that our Unified SuperStandard Model, at current energies, may be close to the GUT level at which coupling constants are equal.

4.2.8.7.2 UNIFIED SUPERSTANDARD MODEL LAGRANGIAN COUPLING CONSTANTS

We begin with the Unified SuperStandard Model lagrangian density $\mathcal{L}_{TE}$ with coupling constants explicitly displayed[93]

$$\mathcal{L}_{TE} = \mathcal{L}_{TE}(g_S, g_W, g_E, g_{DW}, g_{DE}, g_G, g_V, g_S, g_{GR}, g_\Omega)$$

and fields and space-time coordinates not displayed.

In terms of vacuum expectation values as discussed earlier we see we can write[94]

$$\mathcal{L}_{TE} = \mathcal{L}_{TE}(\Phi_1/m_1, \Phi_2/m_2, \ldots, \Phi_{11}/m_{11})$$

where

$$|\,\Phi_1, \Phi_2, \ldots, \Phi_{11}; \Pi_1, \Pi_2, \ldots, \Pi_{11}> = C \prod_{i=1}^{11}\{\exp[[(2\pi)^3 m_i/2]^{1/2}\Phi_i[a_{i2}^\dagger(\mathbf{0},m_i) + a_{i2}(\mathbf{0},m_i)]]\}|0>$$

Assuming all $m_i = \kappa^{-1}$ *we obtain*

$$|\,\Phi_1, \Phi_2, \ldots, \Phi_{11}; \Pi_1, \Pi_2, \ldots, \Pi_{11}> = C \prod_{i=1}^{11}\{\exp[[(2\pi/\kappa)^3/2]^{1/2}\Phi_i'[a_{i2}^\dagger(\mathbf{0}, \kappa^{-1}) + a_{i2}(\mathbf{0}, \kappa^{-1})]]\}|0>$$

---

[91] The closeness of all the values to one is suggestive: The value $\alpha = 1$ (or $e = (4\pi)^{1/2} = 3.54$) was the value found in our calculation in the Johnson, Baker, Willey model of QED. Perhaps a larger calculation along the lines of our paper in massless ElectroWeak theory might yield scaled coupling constant values near unity.

[92] The weakness of the Weak interactions is primarily due to the large masses of the Z and W vector bosons – not the values of their coupling constants g and g'.

[93] The 11th coupling constant g_Θ and the Θ symmetry group are discussed later.

[94] The "vacuum" state |0> above also has factors for the vacuum expectation values used for fields that give masses to fermions and vector bosons as described in Blaha (2015b).

Then we can write

$$\mathcal{L}_{TE} = \mathcal{L}_{TE}(\Phi_1', \Phi_2', \ldots, \Phi_{11}')$$

Setting $m_i = g_{CG} = \kappa^{-1}$ = the Planck mass, simplifies the above expressions and *supports the belief that we are close to the unification of all interactions.* However having particles of such large mass makes them undetectable by accelerators. It also seems too large from the viewpoint of physical intuition. Consequently the above may be the correct expressions with masses perhaps in the TeV range.

We finally note

$$\varphi_{1i}|\ \Phi_1, \Phi_2, \ldots, \Phi_{11}; \Pi_1, \Pi_2, \ldots, \Pi_{11}> = \Phi_{1i}|\ \Phi_1, \Phi_2, \ldots, \Phi_{11}; \Pi_1, \Pi_2, \ldots, \Pi_{11}>$$

4.2.8.7.3 BIG BANG VACUUM

At the origin of the universe – the Big Bang – there was a vacuum state in principle. In our earlier books[95] we showed that the universe existed in an ultra-small, but finite, region for an infinitesimal time before it began an explosive inflationary expansion to become the familiar universe. In this time period there were no infinities – a finite temperature and so on.

Thus it is reasonable to assume one of two possibilities for the above ten coupling constants: 1) they have remained unchanged since the beginning, or 2) they have changed with time.

In this section we note, that if our scaling with the Planck mass κ^{-1} in preceding discussions is correct, then it is reasonable to assume that the vacuum state in the beginning is that defined above with |0> including factors setting fermion and vector boson masses as described in Blaha (2015b).

4.2.8.7.4 EVOLVING/SPACE-TIME DEPENDENT COUPLING CONSTANTS

It is possible that the Unified SuperStandard Model coupling constants evolve with time and may also be spatially varying – different constants in different parts of the universe. Presently there is no decisive evidence for either possibility. In this section we will describe the mechanism to support either or both possibilities.

Consider a classical field (time and spatially varying):

$$\Phi(\mathbf{x}, t) = \int d^3k\ [\alpha(k)f_k(x) + \ \alpha^*(k)f_k{}^*(x)]$$

If we define the coherent vacuum state

$$|\alpha> = C \exp\left\{\int d^3k\ [\alpha(k)a_2{}^\dagger(k) + \alpha^*(k)a_2(k)]\right\}|0>$$

[95] Blaha (2015a) and Blaha (2004).

then

$$\varphi_1(x)|\Phi, \Pi> = \Phi(x)|\Phi, \Pi>$$
$$\pi_1(x)|\Phi, \Pi> = \Pi(x)|\Phi, \Pi>$$

where

$$\varphi_i(\mathbf{x}, t) = \int d^3k\ [a_i(k)f_k(x) + a_i^\dagger(k)f_k{}^*(x)]$$

for i = 1, 2 with

$$f_k(x) = e^{-ik\cdot x}/(2\omega_k(2\pi)^3)^{1/2}$$

where $\omega_k = |\mathbf{k}|$.

The coherent vacuu state $|\alpha>$ has a time and spatially varying vacuum expectation value (classical) field. The above equations can be generalized to the case of the eleven coupling constant vacuum expectation values:[96]

$$|\ \Phi_1, \Phi_2, \ldots, \Phi_{11}; \Pi_1, \Pi_2, \ldots, \Pi_{11}> = C \prod_{i=1}^{11} \exp\left\{\int d^3k\ [\alpha_i(k)a_{2i}{}^\dagger(k) + \alpha_i{}^*(k)a_{2i}(k)]\right\}|0>$$

Then all eleven coupling constant vacuum expectation values are space-time dependent:

$$\varphi_{1i}(x)\ |\ \Phi_1, \Phi_2, \ldots, \Phi_{11}; \Pi_1, \Pi_2, \ldots, \Pi_{11}> = \Phi_i(x)\ |\ \Phi_1, \Phi_2, \ldots, \Phi_{11}; \Pi_1, \Pi_2, \ldots, \Pi_{11}>$$

and the Unified SuperStandard Model lagrangian becomes

$$\mathcal{L}_{TE} = \mathcal{L}_{TE}(\Phi_1(x), \Phi_2(x), \ldots, \Phi_{11}(x))$$

for matrix elements between the vacuum defined by $|\alpha>$ and its conjugate,

Thus our formalism can accommodate space-time varying coupling constants should they be found in the Cosmos.

4.2.8.7.5 A Unified SuperStandard Model Lagrangian Without Any Constants

The preceding chapters put coupling constants within the same framework as particle masses completing the process of eliminating constants from The Unified SuperStandard Model lagrangian. Instead the vacuum contains the values of all coupling constants and particle masses. In one sense this new formulation is a tradeoff. The values of all constants are shifted to the vacuum. However the shift has some advantages technically. One advantage is the ability to have space-time dependent coupling constants as shown above. It would also be straightforward to make masses, and mixing angles, space-time dependent. The possible space-time dependence of coupling constants and particle masses has been an active area of experimental interest for

[96] The "vacuum" state $|0>$ also has factors for the vacuum expectation values used for fields that give masses to fermions and vector bosons as described in Blaha (2015b).

many years although cosmological data seems to indicate these quantities have not changed significantly since the universe began.

The question of changes in lagrangian coupling constant physical values is of great philosophical importance since it appears that the existence of life, as we know it, depends sensitively on their values. This dependence has been embodied in the Anthropomorphic Principle and studied by a number of physicists and philosophers.

Since our formulation allows space-time varying physical constants the question of the Anthropomorphic Principle attains new importance. As we saw in the case of the theory of Black Holes, which was a theory without evidence for over forty years before Black Holes were discovered, Nature seems to provide phenomena that have been shown to be theoretically possible. Many other cases of this sort have also occurred – the most recent example at the time of this writing is Weyl fermions.

Lastly, our approach opens the possibility of a study of all the many constants in The Unified SuperStandard Model lagrangian *on the same footing* rather than in the piecemeal fashion used up to the present. It replaces the scattered hodge-podge of constants in the lagrangian with a centralized location for all constants in the vacuum state permitting a direct study of their interconnection. *The study of the vacuum now becomes of central importance.*

4.2.8.7.6 The Form is Determined But Not the Constants

The derivation of The Unified SuperStandard Model here, and in the Blaha (2015) books and earlier work, was based on Asynchronous Logic (to support parallel physical processes spread in space and time); on complex space-time coordinates, the Complex Lorentz group and complex General Coordinate transformations; the Reality group to map complex coordinates to the real-valued coordinates that we observe; the Generations group that yields the four fermion generations and particle number interactions such as the baryon number and lepton number interactions, and the Species group for complex general coordinate transformations.

This firm basis in fundamental considerations enables us to forge a path to The Unified SuperStandard Model, which included the known features of The Standard Model. We thus were able to avoid the many possible variants and extensions of The Standard Model that have been considered in the Physics literature over the past thirty years.

Two remarkable features of The Unified SuperStandard Model derivation were:

1. A precise fixing of the form of the Unified SuperStandard Model.
2. The absence of any constraints on the values of its coupling constants or masses.

Particle masses were fixed by either the original Higgs Mechanism or by our new mechanism that was based on an extension of Quantum Field Theory to include classical fields such as the vacuum expectation values that cropped up in the original Higgs Mechanism and were handled "by hand." (See Blaha (2015c).)

Thus, up to this point, we have a Unified SuperStandard Model (known) except for a basis for the values of the coupling constants that appear in the theory. The coupling constants have a wide range of values. A fundamental basis for their values has been wanting.

4.2.8.7.7 How Can Coupling Constants be Determined?

The renormalized coupling constants of The Unified SuperStandard Model can be determined experimentally. However a theoretical determination is lacking. This gap in our understanding suggests that there is a major aspect of fundamental physics that is not understood. The fact that we can determine the form – but not the values of coupling constants – so directly from basic principles suggests that a new basic principle(s) is needed to complete The Unified SuperStandard Model. A similar comment applies to fermion and boson masses – both our mechanism,[97] and the vanilla Higgs Mechanism, arbitrarily fixes particle masses. (Attempts to relate particle masses using various symmetries beg the question. As Isidore Rabi (Columbia) once said in a different context, "Who ordered them?" Proposed symmetries are typically "pulled out of a hat.")

The *one* meaningful attempt to determine a coupling constant in a non-trivial 4-dimensional quantum field theory was that of Johnson, Baker and Willey[98] in a 4-dimensional model – massless Quantum Electrodynamics. They developed the theory to the point where if one function, that they called the eigenvalue function, had a zero at the value of the fine structure constant $\alpha \approx 1/137$ then the theory would have no infinities. Adler then showed that the eigenvalue constant zero must be an essential singularity, IF it had a zero,. This author then developed an approximate solution for the eigenvalue function in perhaps the most comprehensive 4-dimensional quantum field theory calculation to all orders in α. The approximate calculation agreed with known exact results to 6th order in e.[99] This author found a zero at $\alpha = 1$. Thus the eigenvalue function zero made the Johnson, Baker, Willey model QED finite removing all divergences.

While the Johnson, Baker, Willey model QED was not successful in finding $\alpha = 1/137$, its method illustrates one possible approach to determining the coupling constants of The Unified SuperStandard Model. It might be possible to use a consistency condition(s) to fix coupling constant values. However, since The Unified SuperStandard Model does not have infinities, the motivation of Johnson, Baker and Willey is absent. A fundamental set of consistency conditions is not apparent and so this approach is not currently viable.

What other approaches are possible? There is an anthropomorphic approach which posits the necessity of certain ranges of some coupling constants for human life, and life in general, to exist. We are not comfortable with this approach since it seems to "beg the question." The input is equivalent to the output mitigating is character as fundamental.

---

[97] Blaha (2015c).

[98] M. Baker and K. Johnson, Phys. Rev. **D8**, 1110 (1973) and references therein.

[99] Equation 1 in our paper S. Blaha, Phys. Rev. **D9**, 2246 (1974).

One could also study the set of coupling constants in a 10-dimensional space looking for the set of values.

Given these considerations we have chosen to pursue a less ambitious approach: to specify the coupling constants as vacuum expectation values of a set of new Higgs-like scalar fields. This approach conceptually parallels the determination of particle masses as vacuum expectation values of scalar Higgs fields.

After reducing coupling constants to vacuum expectation values we considered the possibility that the vacuum state at the Big Bang point determined the coupling constants. We also considered the possibility that coupling constants evolve slowly with time and/or may vary in differing spatial locations.

4.2.9 Status of the Derivation Now

Up to this point in our derivation of the Unified SuperStandard Model we have relied directly on the Complex Lorentz group for the derivation of the form of the fermion spectrum with four fermion species (three up-type and down-type quark subspecies) in one generation,[100] for the derivation of the form of the internal symmetry Reality group, which is identical to that of the Standard Model with the addition of a Dark SU(2)⊗U(1) symmetry and Dark fermions, and for the use of the Higgs Mechanism to give masses to particles. We also developed the formalisms for Two-Tier Field Theory and PseudoQuantum Field Theory.

[100] Generations and layers will be described next.

4.2.10 Generation Group

We now turn to the derivation of the Generation Group and subsequently the Layer Groups, which are based on conserved (and almost conserved) particle numbers such as Baryon Number Conservation.

In this subsection we consider the extension of the one generation theory to four generations based on a new U(4) symmetry group that we call the Generation Group.

It is based on four conservation laws for Baryon, Lepton, Dark Baryon, and Dark Lepton Numbers. In Blaha (2017b) we showed that the existence of a long range baryonic force[101] supports baryon number conservation in a manner similar to electric charge conservation due to the electromagnetic force. Lepton number conservation suggests a very weak long range force as well. By analogy we postulate similar Dark particle number conservation laws and forces.

Having four conserved (or almost conserved) particle numbers with their attendant forces (interactions) leads naturaly to a U(4) symmetry with four 'diagonal' generators that we call the U(4) Generation Group.

If Baryon Number and Lepton Number are both conserved quantities then any linear combination of them is also conserved. Therefore

$$B' = aB + bL$$

is also conserved.

If we consider the Dark Matter sector of the Extended Standard Model it is reasonable to assume that *Dark Baryon Number B_D and Dark Lepton Number L_D are conserved* also (although there is no experimental evidence available as yet to confirm (or deny) these assumptions.)

Thus we have four conserved particle Numbers. Linear combinations of these numbers are also conserved:

$$B' = aB + bL + cB_D + dL_D$$
$$L' = eB + fL + gB_D + hL_D$$
$$B_D' = iB + jL + kB_D + lL_D$$
$$L_D' = mB + nL + oB_D + pL_D$$

The set of 4×4 matrices form an U(4) group if we wish to perform these transformations within lagrangians of the type of the Extended Standard Model. The choice of U(4) rather than SU(4) is required since there are four independent particle Numbers. U(4) has four diagonal matrices

[101] See Blaha (2017b) and earlier books for a discussion of evidence for a baryonic force and conservation law. We found that gravity experiments suggested a possible baryonic potential with the (order of magnitude) coupling constant $\alpha_B = \beta^2/4\pi \simeq .118\ Gm_H^2$ where m_H is the mass of the hydrogen atom.

in its algebra while SU(4) only has three diagonal matrices. U(4) preserves the independence of the four independent particle Numbers. It also allows complex rotations.

The new U(4) Generation Group symmetry leads immediately to four fermion generations. We add an index to each fermion field ranging from 1 through four, add a U(4) gauge field to the covariant derivative,[102] and Yang-Mills local gauge field terms to the lagrangian.

Then we perform the following tasks:[103]

- Define the Two-Tier Lepton and Quark Sectors

- Introduce Symmetry Breaking via the Higgs Mechanism for Fermions and U(4) Gauge Fields

Thus we now have four fermion generations, broken Generation Group symmetry, and masses for fermions and Generation Group gauge fields.

[102] See subsection 4.2.8.4.

[103] These tasks are described in detail in chapters 12 and 13 in Blaha (2017b) as well as earlier books.

4.2.11 Layer Group

This section[104] describes the Layer Groups. The transformations of the Complex Lorentz group transformations led to the internal symmetry Reality group: R = SU(3)⊗SU(2)⊗U(1)⊗SU(2)⊗U(1), which is the symmetry group of the 'original' Standard Model plus Dark factors. The U(4) Generation group was shown to follow from the four particle number conservation laws,[105] and increased the Standard Model group to SU(3)⊗SU(2)⊗U(1)⊗SU(2)⊗U(1)⊗U(4).

We now consider the possibility that there is a further set of conserved particle numbers that lead to replication of the four generations of fermions. We define n replications of the known 4 (possibly 3?) generation fermion spectrum. Then it becomes possible to consider *vertical* rotations amongst the n layers for *each* generation. The vertical rotations for generation k can be symbolized by:

$$\begin{matrix} G_k \\ G_k \\ G_k \\ \cdots \\ G_k \end{matrix} \updownarrow$$

These rotations are based on n 'conserved' numbers L_1, L_2, L_3, L_4 that sum the total number of fermions over the n layers in generation i in any given state.[106] *L_i counts the number of fundamental fermions in generation i summed over all the layers for i = 1, 2, 3,4 in any state.*

The total layer number L_i per generation i is the sum of four other 'conserved' layer numbers:

$$L_i = L_{iB} + L_{iDB} + L_{iL} + L_{iDL}$$

where

L_{iB} – The Bayon layer number
L_{iDB} – The Dark Baryon layer number
L_{iL} – The Lepton layer number
L_{iDL} – The Dark Lepton layer number

for i = 1, 2, 3, 4.

Thus each generation has four 'conserved' layer particle numbers. Since we can perform complex rotations among the n layers of generation i, and since there are four diagonal number operators for each generation, consistency implies that there must be n = 4 layers and each generation must have an associated U(4) group.

---

[104] Much of this chapter appears in Blaha (2017c).
[105] Baryon number, Lepton number, and their Dark analogues.
[106] The generations are numbered from 1 to 4 with the lowest masses generation (e, ν_e, u, d), and its Dark analogues, being generation 1.

These considerations justify our definition of four U(4) Layer groups – one for *each* of the four generations yielding a total Layers group $U(4)^4$. *Each generation in the four layers has a U(4) transformation symmetry group mixing the fermions in its generation across all four layers.*

Fermions have positive $L_i = +1$ values and anti-fermions have negative $L_i = -1$ values. For example, if a state has 3 u quarks, 1 d quark, 1 anti-s quark, 2 electrons, 2 anti-τ leptons, 2 Dark muon neutrinos, and one Dark electron neutrino ν_{De} then $L_1 = 3+1+2+1 = 7$, $L_2 = -1+2 = 1$, $L_3 = -2$ and $L_4 = 0$.

The rationale for the new Layer groups is similar to that of the Generation group which was based on the four particle numbers B, L, B_D, and L_D. The Generation group was based on U(4) rotations related to the four number operators B, L, B_D, and L_D in the one generation SuperStandard Model.

Figure 4.3. The four layers of fermions. Each layer (oval) has four generations of fermions. Layer 1 is the layer that we have found experimentally. The 4th generation of layer 1, the Dark part of layer 1, and the remaining three more massive layers constitute Dark matter.

It is important to note that the Layer particle numbers are independent of the baryon and lepton particle numbers that form the basis of the Generation group, and so the physics embodied in the Generation group is not the same as the physics of the Layer groups defined here.

Layer numbers are conserved under strong and electromagnetic interactions but broken by the Electromagnetic and Weak interactions as well as their Dark counterparts.

Since the coefficients in Layer group transformations can be local functions, the new Layer groups are implemented with Yang-Mills fields.

Just as we extended the reach of the Generation group from one generation to four generations, we extend each of the Layer groups representations to 4 similarly by assuming that there are four layers of fermions for each generation. We add a U(4) layer index to each fermion field for each of the four Layer groups. Fig. 4.4 shows the four U(4) Layer groups rotations of each of the four generation's fermions.

Further the gauge fields for SU(3)⊗SU(2)⊗U(1)⊗SU(2)⊗U(1)⊗U(4) must now be different for each layer.Thus interactions of these types between fermions in different layers is prevented.

To implement this new Layer symmetry we expand the SuperStandard Model lagrangian with the following steps:

1, All covariant derivatives must expand to four U(4) Layer gauge fields terms. The new terms are for four layers of SU(3)⊗SU(2)⊗U(1)⊗SU(2)⊗U(1)⊗U(4) fields. Each gauge field then has an additionl index specifying its layer. The rationale for the choice of these sets of groups to be duplicated is that they, and only they, all play a necessary role in determining the structure of the Periodic Table of Fermions.

2. The new Layer groups are the same as in the Unified SuperStandard Model of Blaha (2017b). Each fermion particle field has an index labeling the layer of the particle making four layers of four generations of fermions.

3. Each layer should have its own set of Higgs particles (modulo mixing) contributing to fermion masses. A layer index number must be added to each Higgs field. One expects that the masses of fermions should be substantially larger for the three 'upper' layers beyond our layer.[107] Otherwise we would have found particles from these upper layers.

The form of the "periodic tables" of fermion and vector bosons that results appears in Figs. 4.4 and 4.5. The 'splitting' of a single fermion generation into four generations and then into four layers is shown in Fig. 4.6.

[107] The interplayed mixing of the particles in each generation between different layers may partly explain the vast increases seen in fermion masses as one goes from generation to generation in each species. The Higgs particles in different layers are different and a possible source of the growth of fermion masses.

THE FERMION PERIODIC TABLE

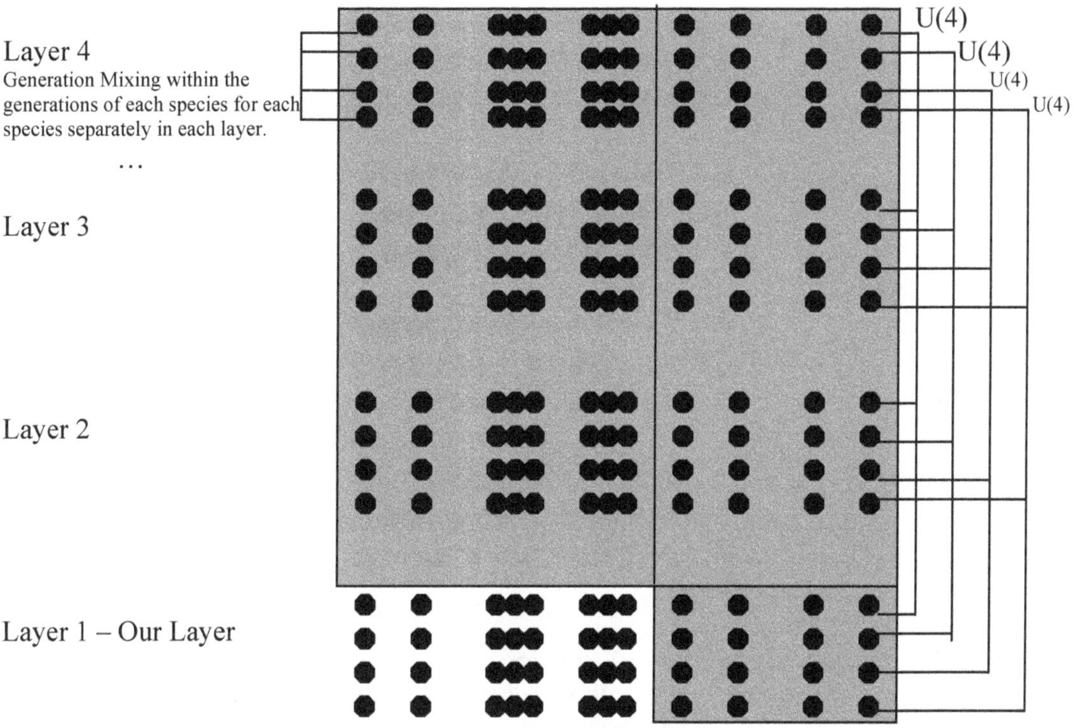

Figure 4.4. Partial example of pattern of mass mixing of the Generation group and of the Layer groups.. Dark parts of the periodic table are gray. Light parts are the known fermions with an additional, as yet not found, 4th generation shown. The lines on the left side show an example of the Generation mixing within one species. The Generation mixing applies to each species in each layer. The lines on the right side show the Layer mixing generation by generation.among all four layers for each generation individually.

THE VECTOR BOSON PERIODIC TABLE

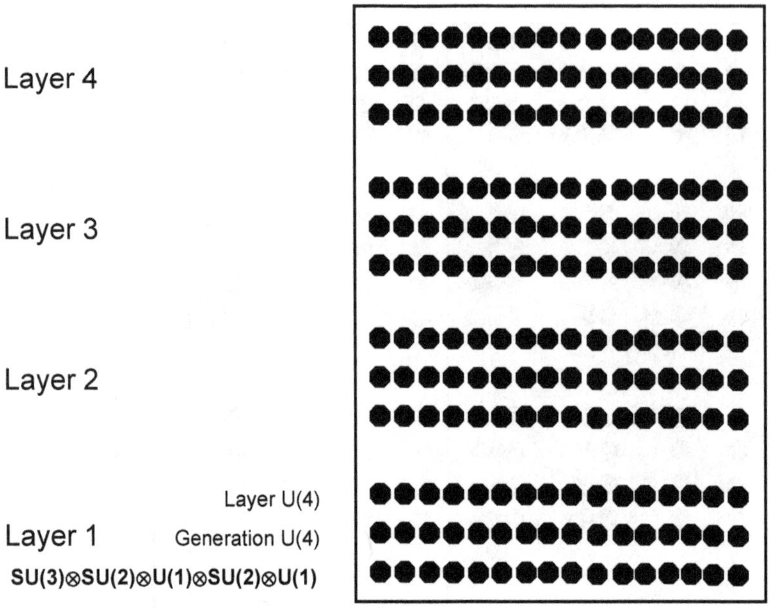

Figure 4.5. The known vector bosons are in the lowest row. The Layer groups are distributed by layer symbolically although they each straddle all four layers.

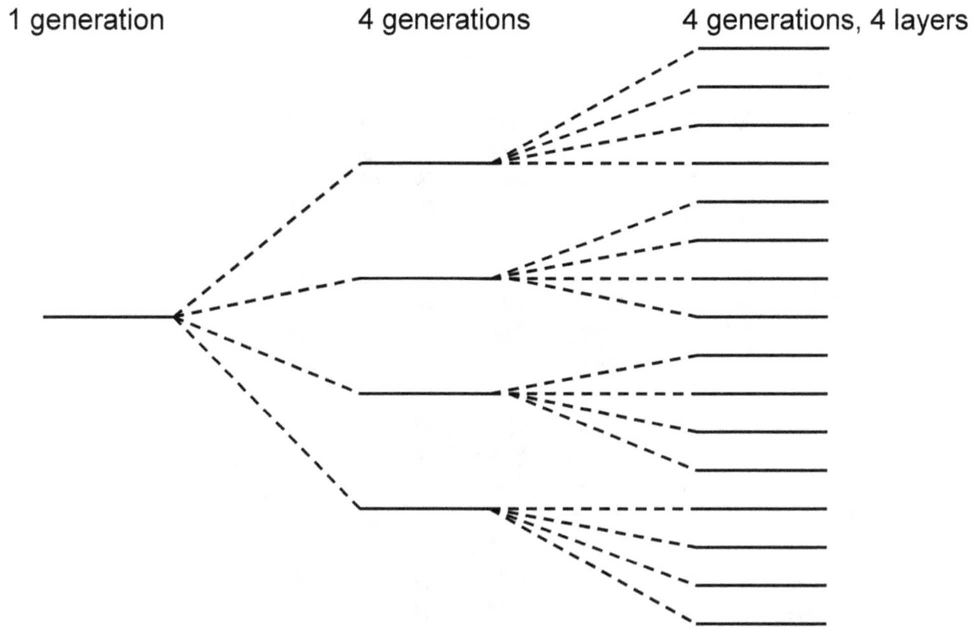

Figure 4.6. The 'splitting of a single generation fermion into four generations and then into four layers.

4.2.11.1 New Labeling of Fermion Periodic Table Particles

Fig. 4.7 has the Periodic Table of Fermions with rows and columns labeled with quadruplets of numbers. In the case of normal quark species, which each consist of a color triplet, a fifth integer would be needed to identify each color quark within a triplet: (T, S, L, G, C) where C = 1, 2, 3 identifies the color ("red, white or blue"). For example, the second generation, fourth layer, normal up-type red quark is (1, 3, 4, 2, 1) where we treat 'red' as having the value 1.

The quadruplet numbering patern is as follows. Species are numbered from 1 through 4 – separately for normal and Dark species: charged lepton, neutral lepton, three up-type quarks, three down-type quarks, Dark charged lepton, Dark neutral lepton, one Dark up-type quark species, and one Dark down-type quark species. Layers are numbered from 1 through 4 with our layer being layer 1. Generations are numbered from 1 through 4 from lightest to the heaviest.[108] (e, ν_e, u and d constitute the known part of generation 1 of our layer.) We will call the quadruplet (or quintet) of a fermion its *ID number*. See Fig. 4.7.

[108] The ordering by mass may not hold in the currently Dark part of the fermion spectrum.

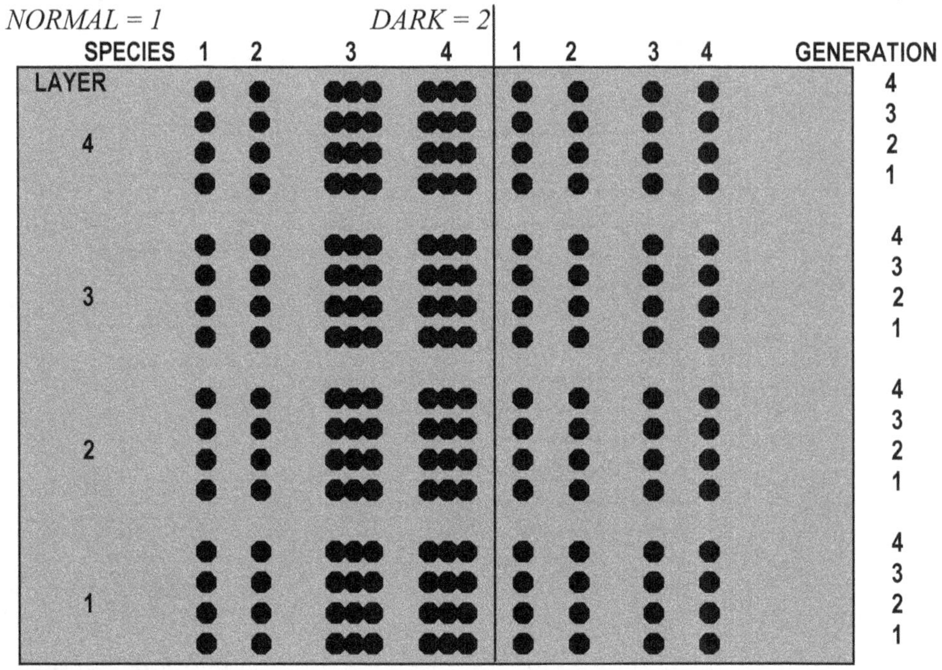

Figure 4.7. The Periodic Table of Fermions with each fermion identified by a quadruplet of integers: (T, S, L, G) where T identifies Normal or Dark, S identifies the species, G identifies the generation, and L identifies the layer. For example the electron has the triplet (1, 1, 1, 1), and the second generation, fourth layer, Dark up-type quark is (2, 3, 4, 2).

Since the Periodic table of Fermions is 2-dimensional like the Chemical Periodic Table of Elements one might ask: Why not use two numbers to identify fermions in a manner similar to the Chemistry table? The answer is related to the group structure of the Extended Standatd Model interactions. Except for T (the type of matter), the other three digits in a quadruplet are each related to a group. Thus G specifies the generation and thus the position of a fermion in the 4 of the U(4) Generation group. The integer L specifies the layer and thus the position of the fermion in the 4 of the form of the U(4) Layer group.

The integer S specifies the position of a fermion amongst the species. In Blaha (2017b) we saw that the four species of each type of matter follow from a U(4) group, that we called the Species Group which was derived from a consideration of the structure of Complex General Coordinate transformations. The U(4) Species gauge fields are denoted $A_S{}^{\mu}(x)$ in discussions below.

Thus the fermion quadruplet labeling is physically motivated by the group structure of the SuperStandard Model: $[SU(3)\otimes SU(2)\otimes U(1)\otimes SU(2)\otimes U(1)\otimes U(4)\otimes U(4)]^4\otimes U(192)$*, which we will explain in detail later.*

4.2.12 Equipartition Principle for Fermions and Gauge Fields

We now[109] will suggest a rationale for the dominant abundance of Dark mass-energy:

4.2.12.1 Equipartition Principle for Particle Degrees of Freedom

In a closed system at equilibrium the thermal energy of a system is equally partitioned (distributed) among its degrees of freedom. This Equipartition Principle is well known. The application of this principle to the beginning of the universe *when all particles were massless* and all symmetries were unbroken suggests that the distribution of mass-energy should be the same for all degrees of freedom at that time. Thus there should be approximately equal numbers and energies of 192 fermions and 192 vector bosons with the same fraction of the total thermal energy.

We now estimate the relative proportion of Normal and Dark matter in the universe at its beginning based on this Equipartition Principle.

4.2.12.2 Proportion of Dark Mass-Energy in the Universe

First we note that 8 of the 12 fermion species (counting quark of each color as a separate species) in layer 1 – the layer with which we are familiar are Normal matter fermions. (Our discussion is based on our SuperStandard Model.) Four of the 12 first layer species are Dark.

The other three fermion layers are all Dark from our point of view since they have not been detected. Thus we find that 40 of the 48 species are Dark yielding a *percentage of Dark Matter equal to 40/48 = 83.33%*. (The same counting could have been done by counting individual fermions with the same results.)

Recent studies of the proportion of Dark Matter in the universe have yielded two estimates: 84.5% by Aghanim et al in Astronomy and Astrophysics 1303;5062 and 81.5% from a NASA fit to various models.

Thus our estimate based on our fermion Equipartition Principle is midway between these experimental estimates.

Two possibilities emerge with respect to the present proportion of Dark Matter:

1. The percentage has not changed from the Beginning and the approximate estimates are slightly off. The lack of change could be due to the extremely small decay rates of the fermions in the higher layers.

2. The percentage of matter in the upper layers has decreased due to decay and so the current proportion may be somewhat below 83.33%.

---

[109] Some of the material in this section appeared in Blaha (2016a).

4.2.12.3 Proportion of Dark Mass-Energy in the Universe

We know of 12 of the 192 vector bosons in the SuperStandard Model and 24 fermions. Thus we find a total of 348 out of 384 particles are Dark yielding a Dark mass-energy of 91% of the universes mass-energy at the beginning of the universe.

The sum of Dark energy in the universe currently has been estimated to be 68% of the total energy and the energy of Dark Matter is estimated to be 26.8%. The total is 95% - a value that compares favorably with our above approximate estimate of 91%. The dfference in these values can be attributed to various factors. One distinct possibility is the decay of Dark mass-energy from higher layers to the known layer in the 13.8 billion years since the Big Bang.

4.2.13 The 'Interaction Rotations' Θ Group

We now define a new 'Interaction Rotation' group, the Θ-Symmetry group G_Θ. It rotates the 192 gauge field components of

$$\Theta = \prod_{k=1}^{4} SU(3)_k \otimes SU(2)_k \otimes U(1)_k \otimes SU(2)_k \otimes U(1)_k \otimes U(4)_k \otimes U(4)_k$$

where k labels the layer for all factors except the last U(4) factor where it enumerates the four Layer groups defined previously[110]. We can abbreviate Θ as

$$\Theta = [SU(3) \otimes SU(2) \otimes U(1) \otimes SU(2) \otimes U(1) \otimes U(4) \otimes U(4)]^4$$

Each layer has its own set of Standard Model-like gauge fields plus there are four Layer group gauge fields that operate between layers. Thus there is no direct leakage between layers except for the 'Interaction Rotation' group interactions G_Θ which rotate all the components of all of the fields in Θ and the Layer group fields that 'rotate' between layers.. Since G_Θ performs complex-valued transformations of the 192 components of the gauge fields of Θ, we see that G_Θ = U(192).

The 192 gauge fields that are rotated, using vector notation, are

$$\mathbf{A}_I^{\mu} = \Sigma_{a,j}\, {}^a\mathbf{A}_{Ij}^{\mu}(x) \cdot {}^a\mathbf{T}_{Ij}$$

for a = 1,2, 3, 4 which specifies the layer, where j which labels the interactions in ${}^a\mathbf{A}_I^{\mu}$,[111] where ${}^a\mathbf{T}_I$ is specified below and where

$${}^a\mathbf{A}_I^{\mu} = (g^a_1\, {}^a\mathbf{A}_{SU(3)}^{\mu}(x_C),\, g^a_2\, {}^a\mathbf{W}^{\mu}(x)\, ,\, g^a_3\, {}^a\mathbf{A}_E^{\mu}(x),\, g^a_4\, {}^a\mathbf{W}_D^{\mu}(x),\, g^a_5\, {}^a\mathbf{A}_{DE}^{\mu}(x),\, g^a_6\, {}^a\mathbf{U}^{\mu}(x)\, ,\, g^a_7\, {}^a\mathbf{V}^{\mu}(x))$$

Each vector ${}^a\mathbf{A}_I^{\mu}$ is a vector of the gauge fields of the respective interactions in layer a. For simplicity we label the respective layer dependent coupling constants as g^a_1, g^a_2, … , g^a_7. The subscript 'D' labels Dark matter interactions, 'W' labels Weak fields, 'E' labels Electromagnetic fields, $U^{\mu}(x)$ labels U(4) Generation group fields, and $V^{\mu}(x)$ labels U(4) Layer group fields.

Similarly we define a 7-vector of 48 matrix generators for each a:

$${}^a\mathbf{T}_I = ({}^a\mathbf{T}_{SU(3)},\, {}^a\boldsymbol{\tau}_{SU(2)},\, {}^a\mathbf{I}_{U(1)},\, {}^a\boldsymbol{\tau}_{DSU(2)},\, {}^a\mathbf{I}_{DU(1)},\, {}^a\mathbf{G}_{U(4)}\,),\, {}^a\mathbf{G}_{LU(4)})$$

---

[110] These groups appear in the Θ-symmetry group because they are all directly connected to defining the form of the Periodic Table of Fermions. Other groups do not specify aspects of the Periodic Table form.

[111] In the case of the Layer groups' gauge fields ${}^a\mathbf{V}^{\mu}(x)$ the index 'a' merely numbers the four Layer groups since the groups straddle all four layers.

We put a layer index on these matrices: since, although they have the same form in all layers, the respective matrices operate in 'different spaces' for each layer since the gauge fields, and the fermion fields they operate on, differ from layer to layer.

The plenitude of interactions that we have identified and summarized in chapter 25 of Blaha (2017b) leads us to consider the possibility of unification based partly on the 'rotation of interactions' and more fully, on an interactions unification based on the Riemann-Christoffel tensor.[112]

When we use the PseudoQuantum (two fields per gauge particle) formalism[113] the part of these gauge fields interactions within a covariant derivative has the form

$$ {}^{a}\mathbf{A}_I{}^{1\mu}(x)\cdot{}^{a}\mathbf{T}_I + {}^{a}\mathbf{A}_I{}^{2\mu}(x)\cdot{}^{a}\mathbf{T}_I = {}^{a}\mathbf{A}_I{}^{1\mu}{}_k(x){}^{a}\mathbf{T}_{Ik} + {}^{a}\mathbf{A}_I{}^{2\mu}{}_k(x){}^{a}\mathbf{T}_{Ik} $$

summed over a and k.

4.2.13.1 Θ-Symmetry: 'Interaction Rotations' for Fermions

The Θ-Symmetry gauge fields appear in Dirac-like equation covariant derivatives.We define a column vector ψ containing the 4-spinors of all 192 normal and Dark fundamental fermions in our theory based on four generations in four layers as detailed earlier and in Blaha (2016a), (2016b) and (2016c). Then the flat space-time Dirac equation[114] has the form:

$$ \gamma_\mu D^\mu\psi = \gamma_\mu\{\partial^\mu + i\,[g_\Theta A_\Theta{}^{1\mu}(x) + g_\Theta A_\Theta{}^{2\mu}(x) + g_S A_S{}^{1\mu}(x) + g_S A_S{}^{2\mu}(x) + \Sigma_a({}^{a}\mathbf{A}_I{}^{1\mu}(x)\cdot{}^{a}\mathbf{T}_I + $$
$$ + {}^{a}\mathbf{A}_I{}^{2\mu}(x)\cdot{}^{a}\mathbf{T}_I)]\}\psi = 0 $$

with the Θ-Symmetry gauge field terms $A_\Theta{}^{1\mu}(x)$ and $A_\Theta{}^{2\mu}(x)$ inserted. Each of these fields consists of a gauge field multiplied by a U(192) matrix generator in the 192 representation since there are 192 fermion fields.

We now define

$$ \mathbf{A}_I{}^{i\mu} = \Sigma_a\,{}^{a}\mathbf{A}_I{}^{i\mu}(x)\cdot{}^{a}\mathbf{T}_I $$

for i = 1, 2 since we will be performing Θ transformations that will mix all the field components within $\mathbf{A}_I{}^{i\mu}$. Then the Dirac equation becomes

$$ \gamma_\mu D^\mu\psi = \gamma_\mu\{\partial^\mu + i\,[g_\Theta A_\Theta{}^{1\mu}(x) + g_\Theta A_\Theta{}^{2\mu}(x) + g_S A_S{}^{1\mu}(x) + g_S A_S{}^{2\mu}(x) + $$
$$ + b\mathbf{A}_I{}^{1\mu}(x) + \mathbf{A}_I{}^{2\mu}(x)]\}\psi = 0 $$

---

[112] Most of the equations in this section first appeared in Blaha (2017a).
[113] See Blaha (2017b).
[114] Similar considerations apply to other Dirac-like equations described in Blaha (2017b).

The $\mathbf{A}_\Theta{}^{1\mu}(x)$ and $\mathbf{A}_\Theta{}^{2\mu}(x)$ gauge fields each separately have a 192×192 matrix representation:

$$\mathbf{A}_\Theta{}^{i\mu}(x) = \Sigma_n \mathbf{A}_\Theta{}^{i\mu}{}_n(x) T_{U(192)n}$$

for i = 1., 2 where n sums over the 192^2 generators of U(192).

The fields in $A_I{}^{1\mu}(x)$, and $A_I{}^{2\mu}(x)$ separately transform under a Θ-transformation. Under a Θ-transformation we find the Dirac equation transforms to

$$\gamma_\mu\{\partial^\mu + i\,[g_\Theta A'_\Theta{}^{1\mu}(x) + g_\Theta A'_\Theta{}^{2\mu}(x) + + g_S A'_S{}^{1\mu}(x) + g_S A'_S{}^{2\mu}(x) + \\ + \mathbf{A'}_I{}^{1\mu}(x) + \mathbf{A'}_I{}^{2\mu}(x)]\}\psi' = 0$$

where

$$A'_\Theta{}^{1\mu}(x) = C_\Theta(x) A_\Theta{}^{1\mu}(x) C_\Theta{}^{-1}(x) - i\, C_\Theta(x)\partial^\mu C_\Theta{}^{-1}(x)/g_\Theta$$
$$A'_\Theta{}^{2\mu}(x) = C_\Theta(x) A_\Theta{}^{2\mu}(x) C_\Theta{}^{-1}(x)$$
$$\mathbf{A'}_S{}^{1\mu}(x) = C_\Theta(x)\mathbf{A}_S{}^{1\mu}(x)\, C_\Theta{}^{-1}(x)$$
$$\mathbf{A'}_S{}^{1\mu}(x) = C_\Theta(x)\mathbf{A}_S{}^{1\mu}(x)\, C_\Theta{}^{-1}(x)$$
$$\mathbf{A'}_I{}^{1\mu}(x) = C_\Theta(x)\mathbf{A}_I{}^{1\mu}(x)\, C_\Theta{}^{-1}(x)$$
$$\mathbf{A'}_I{}^{2\mu}(x) = C_\Theta(x)\mathbf{A}_I{}^{2\mu}(x)\, C_\Theta{}^{-1}(x)$$
$$\psi' = C_\Theta(x)\psi$$

The effect of the Θ-transformation $C_\Theta(x)$ is to rotate the gauge fields components. It is accompanied by a rotation of the generator matrices components. Together they define an equivalent formulation of the original Dirac equation and thus the fermion sector. The next section provides an explicit simple example of Θ-transformations – an ElectroWeak theory.

Note that a Θ-transformation causes a change of gauge in the $A_\Theta{}^{1\mu}(x)$ field. The other gauge fields, $A_\Theta{}^{2\mu}(x)$, $\mathbf{A}_I{}^{1\mu}(x)$ and $\mathbf{A}_I{}^{2\mu}(x)$, are 'rotated' but do not undergo a change of gauge. (Each of the 48 gauge fields within $\mathbf{A}_I{}^{1\mu}(x)$ and $\mathbf{A}_I{}^{2\mu}(x)$ do undergo their own particular changes of gauge for their own transformation groups.)

4.2.13.2 Broken Θ-Symmetry

Θ-symmetry is broken by the complexon nature of color SU(3) gauge fields and quarks. The color SU(3) gauge fields $\mathbf{A}_{SU(3)}{}^{j\mu}(x)$ for j = 1,2 appearing in $\mathbf{A}_I{}^{i\mu}(x)$ are complexon fields in our formulation of SuperStandard Model. Thus they are functions of complex spatial coordinates

$$x_c = (t, \mathbf{x}_r - i\mathbf{x}_i)$$

Therefore color SU(3) does not admit of Θ-transformations. Consequently Θ-Symmetry is broken to $[SU(3)_{Color}]^4 \otimes U(160)$. This symmetry breaking reduces the number of Θ-Symmetry generators to 160^2.

The symmetry breaking mechanism may be expected to lead to large A_Θ gauge field masses. Together with the likely smallness of the g_Θ coupling constant, this leads us to expect that the A_Θ interaction, which occurs between all 192 fermions, will not be detectable.

4.2.13.3 Example: ElectroWeak-like Theory

ElectroWeak model is an example of a global Θ-transformation which does not include the full gamut of features outlined above. Focussing on the Weak and Electromagnetic interactions we consider the covariant derivative

$$\{\partial^\mu + i\,[g\mathbf{W}^\mu\cdot\boldsymbol{\tau} + g' W_0{}^\mu \tau_0]\}\psi = 0$$

Rotating $W_3{}^\mu$ and $W_0{}^\mu$ with $C^{-1}{}_\Theta$

$$\begin{bmatrix} gW_3{}^\mu \\ g'W_0{}^\mu \end{bmatrix} \rightarrow \begin{bmatrix} g\cos\theta\ Z^\mu + g\sin\theta\ A^\mu \\ -g'\sin\theta\ Z^\mu + g'\cos\theta\ A^\mu \end{bmatrix}$$

and rotating the generator matrices by C_Θ

$$\begin{bmatrix} \tau_3 \\ \tau_0 \end{bmatrix} \rightarrow \begin{bmatrix} \cos\theta\ \tau_3\ - \sin\theta\ \tau_0 \\ \sin\theta\ \tau_3\ + \cos\theta\ \tau_0 \end{bmatrix}$$

and making an appropriate choice of θ_W yields the electromagnetic field A and Z we find

$$g'W_0{}^\mu\ \tau_0 + gW_3{}^\mu\tau_3 \rightarrow Z^\mu[(g\cos^2\theta - g'\sin^2\theta)\tau_3 - \tfrac{1}{2}\sin(2\theta)(g + g')\tau_0 + A^\mu[\tfrac{1}{2}\sin(2\theta)(g + g')\tau_3 + $$
$$+ (g'\cos^2\theta - g\sin^2\theta)\tau_0]$$

If we choose the coefficient of A^μ be e, then

$$e(\tau_3 + \tau_0) = [\tfrac{1}{2}\sin(2\theta)(g + g')\tau_3 + (g'\cos^2\theta - g\sin^2\theta)\tau_0]$$

and thus

$$\tfrac{1}{2}\sin(2\theta)(g + g') = (g'\cos^2\theta - g\sin^2\theta) = e$$

Consequently

$$g' = -(\tfrac{1}{2}\sin(2\theta) - \sin^2\theta)/(\tfrac{1}{2}\sin(2\theta) - \cos^2\theta)$$

If we further choose g = g' (since equal coupling constants is an often stated goal of theorists) for simplicity we find the angle $\theta = \pi/8$ or 22.5° while the usual Weinberg angle θ_W is about 30°. Thus we find a close similarity to standard ElectroWeak theory.[115]

We conclude that ElectroWeak theory can be viewed as an example of a Θ-transformation.[116]

4.2.13.4 Θ-Symmetry and Higgs Fields

The Higgs boson fields η (or our PseudoQuantum alternative) also participate in Θ-symmetry rotations. Consider a composite boson field constructed from the concatenation of all Higgs fields. The term generating the gauge field masses has the form

$$(D^\mu\eta)^\dagger D^\mu\eta$$

Upon rotating gauge fields, mass terms will also correspondingly 'rotate' to maintain the covariance of the overall theory.

The A_Θ gauge fields will also all acquire masses since there is no evidence of long range A_Θ fields experimentally.

4.2.13.5 Path Integral Formulation, and the Faddeev-Popov Method

The path integral formulation of our theory, with the complete set of eleven integrations described here and in Blaha (2017b), is fairly straightforward with one exception. Since we use complex valued coordinates for the color SU(3) gauge theory a somewhat different approach must be followed for it. We detail this approach in Blaha (2015a).

The Faddeev-Popov path integral formulation is the same as that of the $A_\Omega{}^{1\mu}(x)$ path integral formulation presented in Blaha (2017b) with only minor notational differences.

---

115 We can, of course, make the angles equal by adjusting the values of g ang g'.

116 A rotation of fermions was not required in this case because the interactions are not charge changing. In the general case a rotation of interactions also requires a rotation of the 192 fermions.

4.2.14 Complex General Relativity Reformulated

We have seen that Complex Special Relativity is the basis of flat space-time phenomena. Flat space-time coordinates are complex-valued in general. The real-valued coordinates that we experience in everyday life are the result of our measuring instruments: clocks and rulers. Real-valued coordinates are generated from complex-valued coordinates by Reality group transformations.

If flat space-time is governed by Complex Special Relativity then it is clear that curved 'space-time' is governed by Complex General Relativity. Here again there is a Reality group the General Relativistic Reality group – a U(4) group – that maps complex-valued General Relativity coordinates to real-valued curved coodinates.[117] There is a corresponding U(4) internal symmetry reality group that we call the *Species group*. This group rotates fermions within a species, or to a different species, or to a superposition of species (remembering that there are four fermion species.)

We can isolate the General Relativistic Reality group by factoring complex General Relativistic coordinate transformations into parts that consist of a real-valued General Coordinate transformation and complex-valued coordinate transformations. It will be apparent that the General Relativistic Reality group emerges in this discussion. The Species group is distinct from the internal symmetry Reality group of the Standard Model: R = SU(3)⊗SU(2)⊗U(1)⊗SU(2)⊗U(1). We begin by defining the tetrad notation.

4.2.14.1 Tetrad (Vierbein) Formalism

The *vierbein* formalism begins with the Equivalence Principle that allows us to define an inertial coordinate system in the neighborhood of any point Z in space-time. We will use the notation $\varsigma^\alpha(Z)$ to denote the inertial coordinates at Z. We define a tetrad or vierbein as

$$v^\alpha{}_\mu(x) = (\partial\varsigma^\alpha(x)/\partial x^\mu)_{x=Z}$$

and, in a neighborhood of Z, we can invert the relation between ς and x to define an inverse

$$w^\mu{}_\alpha(x) = (\partial x^\mu(\varsigma)/\partial\varsigma^\alpha)_{x=X}$$

such that

$$w^\mu{}_\alpha(x)v^\alpha{}_\nu(x) = \delta^\mu{}_\nu$$
$$w^\mu{}_\beta(x)v^\alpha{}_\mu(x) = \delta^\alpha{}_\beta$$

In real General Relativity all *tetrads* are real-valued. In Complex General Relativity a *tetrad* $v^\alpha{}_\mu(x)$ is complex-valued.

---

[117] Much of this chapter appears in Blaha (2016h) and (2017a).

The metric at a curved space-time point X is defined in terms of *tetrads* as

$$g_{\rho\sigma}(x) = \eta_{\alpha\beta}\, v^{\alpha}{}_{\rho}(x) v^{\beta}{}_{\sigma}(x)$$
$$g^{\rho\sigma}(x) = \eta^{\alpha\beta}\, w^{\rho}{}_{\alpha}(x) w^{\sigma}{}_{\beta}(x)$$

The inverse of a *tetrad* transformation can also be expressed as

$$w_{\beta}{}^{\nu}(x) = v_{\beta}{}^{\nu}(x) = \eta_{\beta\alpha} g^{\nu\mu}(x) v^{\alpha}{}_{\mu}(x)$$

Then a *tetrad* and its inverse satisfy the relations

$$v^{\alpha}{}_{\mu}(x) v_{\beta}{}^{\mu}(x) = \delta^{\alpha}{}_{\beta}$$

and

$$v^{\alpha}{}_{\mu}(x) v_{\alpha}{}^{\nu}(x) = \delta^{\nu}{}_{\mu}$$

There are two general types of space-time transformations that can be performed on a tetrad.

1. A complex-valued (possibly real-valued) General Relativistic coordinate transformation:

$$v'^{\alpha}{}_{\mu}(x) = \partial x^{\nu}/\partial x'^{\mu}\, v^{\alpha}{}_{\nu}(x)$$

2. A complex-valued, local *Lorentzian transformation*

$$v'^{\beta}{}_{\mu}(x) = \Lambda(x)^{\beta}{}_{\alpha}\, v^{\alpha}{}_{\mu}(x)$$

where $\Lambda(x)^{\beta}{}_{\alpha}$ is an element of a subset of the local Complex Lorentz Group.

The local Lorentzian transformations $\Lambda(x)^{\beta}{}_{\alpha}$ consist of local Lorentz transformations that are real-valued, and complex-valued Lorentz transformations. Both types of transformations satisfy the orthogonality condition:

$$\eta_{\alpha\beta}\Lambda^{\alpha}{}_{\rho}(x)\Lambda^{\beta}{}_{\sigma}(x) = \eta_{\rho\sigma}$$

Thus the *tetrad* partakes of both local (position dependent) General Relativistic transformations and local Lorentzian transformations.

4.2.14.2 Complex General Relativistic Transformations

The General Relativistic Reality group interaction emerges from complex General Relativistic transformations. We can separate elements of the set of all complex General

Coordinate transformations into a product of two factors: a real-valued General Coordinate transformation and a complex-valued General Coordinate transformation. The set of complex factors can be further factored into those that satisfy

$$\Lambda(\omega, \mathbf{u})^T G \Lambda(\omega, \mathbf{u}) = G$$

and those that do not. We then see that the set of those that do not satisfy the above equation form a curved space representation of the U(4) group under 'multiplication' of transformations.

The elements of the set of real and complex General Coordinate transformations whose flat complex space-time limit satisfy the above equation form the elements of the Complex Lorentz group.[118]

We thus find the set of all 4-dimensional complex, curved space General coordinate transformations can be visualized as in Fig. 4.8. The next section describes the interplay of the three parts displayed in Fig. 4.8.

4.2.14.3 Structure of Complex General Coordinate Transformations

Complex General Coordinate transformations can be uniquely factored into products of two terms, which will later be further factored into three factors. They have the form

$$\partial x''^{\nu}(x)/\partial x^{\mu} = U(x'')^{\nu}{}_{\beta}\, \partial x'^{\beta}(x)/\partial x^{\mu}$$

where

$$x''^{\nu}(x) = U(x'')^{\nu}{}_{\beta} x'^{\beta}$$
$$x'^{\mu}(x) = U^{-1\mu}{}_{b}(x'')\, x''^{b}$$

where $U(x')^{\nu}{}_{\beta}$ is complex and where $\partial x'^{\beta}(x)/\partial x^{\mu}$ is a purely real General Coordinate transformation.

We define

$$U\,(x'')^{\mu}{}_{\nu} = w^{\mu}{}_{a}(x'')\left[\exp\left(i \textstyle\sum_k g_k \Phi_k(x'')\tau_k\right)\right]^{a}{}_{b}\, v^{b}{}_{\nu}(x'')$$
$$U^{-1}(x'')^{\mu}{}_{\nu} = w^{\mu}{}_{a}(x'')\left[\exp\left(-i \textstyle\sum_k g_k \Phi_k(x'')\tau_k\right)\right]^{a}{}_{b}\, v^{b}{}_{\nu}(x'')$$

where the constants g_k are real, and Φ_k and τ_k are hermitean. The uniqueness of the factorization follows from the Reality group (and U(4)) property that any complex 4-vector can be uniquely mapped to any specified real 4-vector.

---

118 It is this part of curved space-time General Relativity that becomes the flat space-time Complex Lorentz group, which leads to the SU(3)⊗SU(2)⊗U(1)⊗SU(2)⊗U(1) Standard Model Reality group.

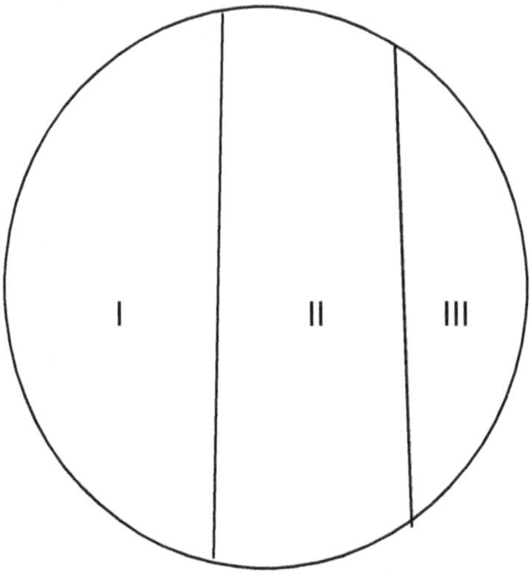

Figure 4.8. A visualization of the set of General Coordinate transformations separated into real-valued General coordinate transformations (part I), complex transformations that satisfy $\Lambda(\omega, \mathbf{u})^T G \Lambda(\omega, \mathbf{u}) = G$ (part II), and complex transformations that do not satisfy $\Lambda(\omega, \mathbf{u})^T G \Lambda(\omega, \mathbf{u}) = G$ (part III). Part I and part II combine in the limit of flat space-time to form the Complex Lorentz group. Parts II and III elements form a U(4) group that we call the General Relativistic Reality group.

Given the factorization above it becomes possible to separate the affine connection correspondingly.

4.2.14.4 Complex Affine Connection – General Relativistic Reality Group

The structure of a complex general coordinate transformation enables us to calculate its affine connection for later use in determining the covariant derivative, and the dynamic equations. First the transformation to the real-valued x' coordinates from inertial coordinates is

$$\Gamma^{\sigma}{}_{\lambda\mu}\,(x') = \partial x'^{\sigma} / \partial \varsigma^{\rho}\ \partial^2 \varsigma^{\rho} / \partial x'^{\lambda} \partial x'^{\mu}$$

Next the Reality group transformation has the affine connection

$$\Gamma^{\sigma}{}_{\lambda\mu}\,(x'') = \partial x''^{\sigma} / \partial \varsigma^{\rho}\ \partial^2 \varsigma^{\rho} / \partial x''^{\lambda} \partial x''^{\mu}$$

which can be re-expressed as

$$\Gamma^{\sigma}{}_{\lambda\mu}(x'') = \partial x''^{\sigma}/\partial x'^{\beta}\, \partial x'^{\beta}(\varsigma)/\partial\varsigma^{\rho}\, \partial/\partial x''^{\mu}[\partial\varsigma^{\rho}/\partial x'^{\alpha}\, \partial x'^{\alpha}/\partial x''^{\lambda}]$$
$$= \partial x''^{\sigma}/\partial x'^{\beta}\, \partial x'^{\alpha}/\partial x''^{\lambda}\, \partial x'^{\gamma}/\partial x''^{\mu}\, \Gamma^{\beta}{}_{\alpha\gamma}(x') + \partial x''^{\sigma}/\partial x'^{\beta}\, \partial^{2} x'^{\beta}/\partial x''^{\lambda}\partial x''^{\mu}$$

Next substituting the General Relativistic Reality group transformation

$$x''^{\nu}(x) = U(x'')^{\nu}{}_{\beta}x'^{\beta}$$
$$x'^{\mu}(x) = U^{-1}(x'')^{\mu}{}_{\beta}\, x''^{\beta}$$

together with

$$\partial x''^{\sigma}/\partial x'^{\beta} = \partial[U(x'')^{\sigma}{}_{\alpha}x'^{\alpha}]/\partial x'^{\beta} = U(x'')^{\sigma}{}_{\beta} + x'^{\alpha}\,\partial U(x'')^{\sigma}{}_{\alpha}/\partial x'^{\beta}$$

$$\partial x'^{\sigma}/\partial x''^{\beta} = \partial[U^{-1}(x'')^{\sigma}{}_{\alpha}x''^{\alpha}]/\partial x''^{\beta} = U^{-1}(x'')^{\sigma}{}_{\beta} + x''^{\alpha}\,\partial U^{-1}(x'')^{\sigma}{}_{\alpha}/\partial x''^{\beta}$$

we find the second term above is the Reality fields affine connection

$$\Gamma_{R}{}^{\sigma}{}_{\lambda\mu}(x'') = \partial[U(x'')^{\sigma}{}_{\alpha}x'^{\alpha}]/\partial x'^{\beta}\, \partial\{\partial[U^{-1}(x'')^{\beta}{}_{\alpha}x''^{\alpha}]/\partial x''^{\lambda}\}/\partial x''^{\mu}$$

and so we find the affine connections are approximately additive. Thus approximately

$$\Gamma^{\sigma}{}_{\lambda\mu}(x'') = \Gamma_{GR}{}^{\sigma}{}_{\lambda\mu}(x') + \Gamma_{R}{}^{\sigma}{}_{\lambda\mu}(x'')$$

if $x''^{\sigma} \simeq x'^{\sigma}$.

A complex transformation of types II and III in Fig. 4.8 has the form:

$$U(x'')^{\mu}{}_{\nu} = w^{\mu}{}_{a}(x'')[\exp(i\textstyle\sum_{k}\Phi_{k}(x'')\tau_{k})]^{a}{}_{b}\, l^{b}{}_{\nu}(x'')$$
$$U^{-1}(x'')^{\mu}{}_{\nu} = w^{\mu}{}_{a}(x'')[\exp(-i\textstyle\sum_{k}\Phi_{k}(x'')\tau_{k})]^{a}{}_{b}\, l^{b}{}_{\nu}(x'')$$

Its infinitesimal transformation is approximately

$$U(x'')^{\nu}{}_{\beta} \approx \delta^{\nu}{}_{\beta} + i\textstyle\sum_{k}\Phi_{k}(x'')[\tau_{k}]^{\nu}{}_{\beta}$$
$$U^{-1}(x'')^{\nu}{}_{\beta} \approx \delta^{\nu}{}_{\beta} - i\textstyle\sum_{k}\Phi_{k}(x'')[\tau_{k}]^{\nu}{}_{\beta}$$

using the *vierbein* flat space-time limits

$$w^{\mu}{}_{a}(x'') \approx \delta^{\mu}{}_{a}$$
$$l^{b}{}_{\nu}(x'') \approx \delta^{b}{}_{\nu}$$

where

$$\Phi_{k}(x) = \int^{x} dy_{\lambda}\, A_{Rk}{}^{\lambda}(y)$$

Then

$$\Gamma_{R}{}^{\sigma}{}_{\lambda\mu} = -\tfrac{1}{2}i\{\textstyle\sum_k A_{Rk}(x'')_{\mu}[\tau_k]^{\sigma}{}_{\lambda} + \sum_k A_{Rk}(x'')_{\lambda}[\tau_k]^{\sigma}{}_{\mu}\}$$
$$= A_R{}^{\sigma}{}_{\mu\lambda} + A_R{}^{\sigma}{}_{\lambda\mu}$$

(summed over k) with the matrix $A_R{}^{\sigma}{}_{\mu\lambda}$ given by

$$A_R{}^{\sigma}{}_{\mu\lambda} = -\tfrac{1}{2}i\textstyle\sum_k A_{Rk\mu}[\tau_k]^{\sigma}{}_{\lambda}$$

with $A_R{}^{\sigma}{}_{\mu\lambda}$ transformable to matrix row and column numbers

$$A_{R_{flat}}{}^{i\mu a}{}_{b} = A_{R_{flat}}{}^{i\mu}[\tau_k]^{\sigma}{}_{\lambda}\delta_{\sigma}{}^{a}\delta^{\lambda}{}_{b}$$

using the flat space-time vierbein values, and so $A_{R_{flat}}{}^{ia}{}_{\mu b}$ may be written in matrix form as

$$A_{R_{flat}}{}^{i}{}_{\mu} = -\tfrac{1}{2}i\textstyle\sum_k A_{R_{flat}}{}^{i}{}_{k\mu}\tau_k$$

In the flat space-time limit $A_{Rk}{}^{\lambda}(y)$ becomes the Species group U(4) gauge field $A_{R_{flat}k}{}^{\lambda}(y)\tau_k$.

4.2.14.5 PseudoQuantization of Affine Connections

Having obtained the form of the general affine connection we now PseudoQuantize them for later use in our unification program. We define

$$R^{1\beta}{}_{\sigma\nu\mu} = \partial_\mu H^{1\beta}{}_{\sigma\nu} - \partial_\nu H^{1\beta}{}_{\sigma\mu} + H^{1\gamma}{}_{\nu\sigma}H^{1\beta}{}_{\gamma\mu} - H^{1\gamma}{}_{\mu\sigma}H^{1\beta}{}_{\gamma\nu}$$
$$R^{2\beta}{}_{\sigma\nu\mu\rho} = \partial_\mu H^{2\beta}{}_{\sigma\nu} - \partial_\nu H^{2\beta}{}_{\sigma\mu} + H^{2\gamma}{}_{\nu\sigma}H^{2\beta}{}_{\gamma\mu} - H^{2\gamma}{}_{\mu\sigma}H^{2\beta}{}_{\gamma\nu} +$$
$$+ H^{1\gamma}{}_{\nu\sigma}H^{2\beta}{}_{\gamma\mu} - H^{1\gamma}{}_{\mu\sigma}H^{2\beta}{}_{\gamma\nu} + H^{2\gamma}{}_{\nu\sigma}H^{1\beta}{}_{\gamma\mu} - H^{2\gamma}{}_{\mu\sigma}H^{1\beta}{}_{\gamma\nu}$$

where

$$H^{\sigma}{}_{\nu\mu} = \Gamma_{GR}{}^{\sigma}{}_{\nu\mu} + \Gamma_{GR}{}^{2\sigma}{}_{\nu\mu} + \Gamma_R{}^{1\sigma}{}_{\nu\mu} + \Gamma_R{}^{2\sigma}{}_{\nu\mu}$$

where $\Gamma_{GR}{}^{\sigma}{}_{\nu\mu}$ and $\Gamma_{GR}{}^{2\sigma}{}_{\nu\mu}$ are affine connections for real-valued General Relativity, and $\Gamma_R{}^{1\sigma}{}_{\nu\mu}$ and $\Gamma_R{}^{2\sigma}{}_{\nu\mu}$ are affine connections for a complex-valued set of transformations embodying a U(4) gauge group that combine with real-valued General Relativistic transformations to yield Complex General Relativistic transformations.

The affine connection is most often viewed as a derived quantity—part of the derivation of the curvature tensor in General Relativity. It is typically derived from manipulations of the metric $g_{\mu\nu}$. However, the affine connection can also be viewed as a set of independent fields that become related to the metric via dynamic equations.

Some years ago A. Einstein and H. Weyl[119] pointed out that the metric and the affine connection should be treated as independent quantities and subject to independent arbitrary infinitesimal variations:

> "In contrast to Einstein's original "metric" conception in terms of the $g_{\nu\mu}$ there was later developed, by Eddington, by Einstein himself, and recently by Schrödinger, an affine field theory operating with the components $\Gamma^{\sigma}{}_{\nu\mu}$ of an affine connection. But in 1925 Einstein also advocated a "mixed" formulation by means of a lagrangian in which both the $g_{\nu\mu}$ and the $\Gamma^{\sigma}{}_{\nu\mu}$ are taken as basic field quantities and submitted to independent arbitrary infinitesimal variations.[120] In certain respects this seems to be the most natural procedure."

Following this approach we have introduced the above affine connections for use in the construction of our unification of the eleven particle interactions.

4.2.14.6 Species Group U(4) Gauge Fields

From the above discussion we see the flat space-time limit of $A_{Rk}{}^{\lambda}(y)$ is a local U(4) gauge field with a corresponding internal symmetry gauge field $A_{Rflatk}{}^{\lambda}(y)$ – the Species group gauge field. The mathematical features of this field is quite similar to to the U(4) Generation group fields. The interaction that appears in covariant derivatives is $g_8\mathbf{A}_{Rflat}{}^{\mu}(x) = g_8\mathbf{A}_{Rflatk}{}^{\mu}(x)\, \mathbf{G}_{RflatU(4)k}$ where k is summed from 1, … , 16.

4.2.14.7 Species Group Covariance

A Species group transformation on a Dirac equation must be covariant. Consider the Dirac equation with $\mathbf{A}_{Rflat}{}^{\mu}(x)$ but with other gauge field terms in the covariant derivative omitted for simplicity:

$$\bar{\psi}(x)[i\gamma_\mu(\partial/\partial x_\mu - ig_8A_{Rflatk}{}^{\mu}(x)G_{RflatU(4)k}) - m]\psi(x) = 0$$

summed over k. If we perform a Species group transformation U on lagrangian terms:

$$\bar{\psi}(x)[i\gamma_\mu(\partial/\partial x_\mu - ig_8A_{Rflatk}{}^{\mu}(x)G_{RflatU(4)k}) - m]U^{-1}U\psi(x)$$

or

$$\bar{\psi}(x)U^{-1}U[iU^{-1}U\gamma_\mu U^{-1}U(\partial/\partial x_\mu - ig_8A_{Rflatk}{}^{\mu}(x)G_{RflatU(4)k}) - m]U^{-1}U\psi(x)$$

we find

[119] H. Weyl, Phys. Rev. **77**, 699 (1950).
[120] A. Einstein, Sitzungsber., Preuss. Akad. Der Wissensch. (1925), p. 414.

$$\bar{\psi}'(x)[i\gamma_\mu{}'U(\partial/\partial x_\mu - ig_8 A_{Rflatk}{}^\mu(x)G_{R_{flat}U(4)k})U^{-1} - m]\psi'(x)$$

where

$$\gamma_\mu{}'(x) = U\gamma_\mu U^{-1}$$

is locally equivalent to a Dirac matrix by Good's Theorem.[121] If we set

$$A'_{Rflat}{}^\mu(x) = -(i/g_8)U[\partial U^{-1}/\partial x^\mu] + UA_{Rflat}{}^\mu(x)U^{-1}$$

then the transformed lagrangian terms are

$$\bar{\psi}'(x)[i\gamma_\mu{}'(x)(\partial/\partial x_\mu - ig_8 A'_{Rflatk}{}^\mu(x)G_{R_{flat}U(4)k}) - m]\psi'(x)$$

and has the same form as the original terms above and thus is covariant. We note the indices of the matrices $G_{R_{flat}U(4)k}$ are spinor indices and so $G_{R_{flat}U(4)k}\gamma_\mu$ has an implicit spinor matrix summation.

The coordinate dependence of $\gamma_\mu{}'(x)$ introduces locality into the Dirac matrix. This locality might be viewed with concern except that an inverse Species group transformation exists that removes the locality. Thus the physical impact of this 'new' locality is eliminated.

4.2.14.8 Physical Role of the Species Group

The physical role of elements of the Species group is to 'rotate' fermion fields within a given species (for fixed Type, generation and layer), and also to 'rotate' a fermion field from one species to a field in another species (again for fixed Type, generation and layer). Thus if we consider the four types of species: charged lepton, neutral lepton, up-type quark, and down-type quark, then a Species U(4) Group transformation 'rotates' any fermion field to another fermion field within its own species, or to a fermion field in a different species. These rotations do not change the Type (normal or Dark), the generation, or the layer of the fermion field[122] – only its species at most.

If a Species group transformation is applied to a free fermion field it will yield a new free fermion field(s) in the same or a diferent species. If this fermion field has a complex energy then the Species group transformation must be followed by a SuperStandard Model internal symmetry Reality group transformation that transforms the fermion field into a field with a real-valued energy since free fundamental particles must have real-valued energies (as we pointed out earlier) due to the fact that they cannot decay. The fermion field so produced, with real-valued energy, must be in one of the four species (six species if you count colored subspecies as distinct) for each type of matter: normal or Dark.

---

[121] R. H. Good, Jr., Rev. Mod. Phys., **27**, 187 (1955).
[122] It also does not change the color of an SU(3) color quark.

4.2.14.9 Spontaneous Symmetry Breaking of the Species Group

We begin by letting $\mathbf{A}_{\mathrm{Rflat}}{}^{\mu}(\mathrm{x}) = \mathbf{A}_{\mathrm{S}}{}^{\mu}(\mathrm{x})$ *and* $\mathbf{G}_{\mathrm{RflatU(4)}} = \mathbf{G}_{\mathrm{S}}$. Symmetry breaking of the Species group and the generation of fermion and $\mathbf{A}_{\mathrm{S}}{}^{\mu}(\mathrm{x})$ gauge field masses is described in Blaha (2017b).

4.2.15 Derivation Status at this Point

We have derived the form of the Fermion and internal symmetry Reality Group from the Complex Lorentz group including the form of the Dark fermion spectrum and Dark ElectroWeak theory. We have derived the Species Group and the form of Real-valued General Coordinate transformations from Complex General Relativity. The Generation Group and fermion generations followed from the four fermion conserved particle number operators: Baryon Conservation, Lepton Conservation and their Dark analogues. The four Layer Groups followed from four conservation laws for each of the four fermion generations. Higgs particles were shown to follow from complex-valued gauge fields excepting the Strong interaction gauge fields. The Higgs Mechanism was used for symmetry breaking and the generation of fermion and vector boson masses. The 'Rotation of Interactions' Θ-Group was used to introduce a note of unification which was made stronger by the use of a Higgs Mechanism to generate gauge field coupling constants from constant parameters with all known parameters of the same order of magnitude suggesting GUT unification.

Fermion dynamic equations followed from the Complex Lorentz group and the requirement of covariance under gauge transformations.

We now turn to deriving the form of the lagrangian terms for gauge fields and General Relativity.

4.2.16 The Riemann-Christoffel Curvature Tensor and Vector Boson Part of Unified SuperStandard Model Lagrangian

4.2.16.1 The Covariant Derivative

The covariant derivative[123] which appears in fermion and gravitation equations uses

$$^a\mathbf{A}_I{}^\mu(x) = (^ag_1{}^a\mathbf{A}_{SU(3)}{}^\mu(x_C),\ ^ag_2{}^a\mathbf{W}^\mu(x)\ ,\ ^ag_3{}^a\mathbf{A}_E{}^\mu(x),\ ^ag_4{}^a\mathbf{W}_D{}^\mu(x),\ ^ag_5{}^a\mathbf{A}_{DE}{}^\mu(x),\ ^ag_6{}^a\mathbf{U}^\mu(x),\ ^ag_7{}^a\mathbf{V}^\mu(x))$$

where a labels the layer, a = 1, 2, 3, 4. We define the sum over a and the the components of the vector $^a\mathbf{A}_I{}^\mu(x)$ labeled with i by

$$\mathbf{C}_I{}^\mu(x) = \Sigma_{a,i}\ {}^a\mathbf{A}_{Ii}{}^\mu(x) + g_8\mathbf{A}_S{}^\mu(x) + g_\Theta\mathbf{A}_\Theta{}^\mu(x)$$

For simplicity we label the respective coupling constants in layer a as $^ag_1$, $^ag_2$, …, $^a g_7$. In the equation above: the subscript 'D' labels Dark matter interactions, 'W' labels Weak fields, 'E' labels Electromagnetic fields, $V^\mu(x)$ labels U(4) Generation group fields, and 'V' labels U(4) Layer group fields. A_S labels the U(4) Species Group.

The symmetry is $[SU(3)\otimes SU(2)\otimes U(1)\otimes SU(2)\otimes U(1)\otimes U(4)\otimes U(4)]^4\otimes U(4)$. The number of gauge fields for each of the elements of the vectors is 8. 3, 1, 3, 1, 16, 16 totalling 192 components, plus 16 components for the Species gauge fields. (We segregate the Species fields in Complex General Relativistic transformations.)

Using the above definitions the covariant derivative of a 4-vector is

$$\begin{aligned} D_\nu V_\mu &= (\partial_\nu + iF_\nu)V_\mu - H^\sigma{}_{\nu\mu}V_\sigma \\ &= [g^\sigma{}_\mu\partial_\nu + ig^\sigma{}_\mu F_\nu - H^\sigma{}_{\nu\mu}]V_\sigma \\ &= [g^\sigma{}_\mu\partial_\nu + iD^\sigma{}_{\mu\nu}]V_\sigma \end{aligned}$$

where[124]

$$F^\mu = C_I{}^{1\mu}(x) + \mathbf{C}_I{}^{2\mu}(x) + B^{1\mu} + B^{2\mu}$$

and

$$\begin{aligned} H^\sigma{}_{\nu\mu} &= \Gamma_{GR}{}^\sigma{}_{\nu\mu} + \Gamma_{GR}{}^{2\sigma}{}_{\nu\mu} \\ D^\sigma{}_{\mu\nu} &= g^\sigma{}_\mu F_\nu + iH^\sigma{}_{\nu\mu} \end{aligned}$$

where we have abstracted the complex part of the complex affine connection into the U(4) gauge field $A_S{}^\mu$. $H^\sigma{}_{\nu\mu}$ is the real-valued part of the complex affine connection.

---

123 This section has equations obtained from Blaha (2017d).

124 We will omit the insertion of the coupling constants of $B^{1\mu}$ and $B^{2\mu}$ in the interests of simplifying expressions.

Commutators of the vector fields in F_μ are implicit when the covariant derivative is applied to vectors and tensors such as V_σ.

4.2.16.2 The Curvature Tensor

The curvature tensor applied to a 4-vector is

$$
\begin{aligned}
R'^{\beta}{}_{\sigma\nu\mu}V_\beta = {} & g^{\alpha}{}_{\mu}(\partial_\nu + iF_\nu)g^{\beta}{}_{\sigma}(\partial_\alpha + iF_\alpha)V_\beta - H^{\alpha}{}_{\mu\nu}g^{\beta}{}_{\sigma}(\partial_\alpha + iF_\alpha)V_\beta + \\
& + H^{\alpha}{}_{\mu\nu}H^{\beta}{}_{\sigma\alpha}V_\beta - g^{\alpha}{}_{\mu}(\partial_\nu + iF_\nu)H^{\beta}{}_{\sigma\alpha}V_\beta - H^{\gamma}{}_{\nu\sigma}\{g^{\alpha}{}_{\gamma}(\partial_\mu + iF_\mu)V_\alpha - H^{\alpha}{}_{\gamma\mu}V_\alpha\} - \\
& - \{\mu\leftrightarrow\nu\}
\end{aligned}
$$

$$
= ig^{\beta}{}_{\sigma}(\partial_\nu F_\mu - \partial_\mu F_\nu - i[F_\nu, F_\mu])V_\beta + (\partial_\mu H^{\beta}{}_{\sigma\nu} - \partial_\nu H^{\beta}{}_{\sigma\mu} + H^{\gamma}{}_{\nu\sigma}H^{\beta}{}_{\gamma\mu} - H^{\gamma}{}_{\mu\sigma}H^{\beta}{}_{\gamma\nu})V_\beta
$$

$$
\begin{aligned}
= {} & ig^{\beta}{}_{\sigma}(F_E{}^1{}_{\nu\mu} + F_E{}^2{}_{\nu\mu} + F_W{}^1{}_{\nu\mu} + F_W{}^2{}_{\nu\mu} + F_{DE}{}^1{}_{\nu\mu} + F_{DE}{}^2{}_{\nu\mu} + F_{DW}{}^1{}_{\nu\mu} + F_{DW}{}^2{}_{\nu\mu} + F_{SU(3)}{}^1{}_{\nu\mu} + \\
& + F_{SU(3)}{}^2{}_{\nu\mu} + F_U{}^1{}_{\nu\mu} + F_U{}^2{}_{\nu\mu} + F_V{}^1{}_{\nu\mu} + F_V{}^2{}_{\nu\mu} + F_S{}^1{}_{\nu\mu} + F_S{}^2{}_{\nu\mu} + F_\Theta{}^1{}_{\nu\mu} + F_\Theta{}^2{}_{\nu\mu} + F_B{}^1{}_{\nu\mu} + F_B{}^2{}_{\nu\mu})V_\beta + \\
& + (\partial_\mu H^{\beta}{}_{\sigma\nu} - \partial_\nu H^{\beta}{}_{\sigma\mu} + H^{\gamma}{}_{\nu\sigma}H^{\beta}{}_{\gamma\mu} - H^{\gamma}{}_{\mu\sigma}H^{\beta}{}_{\gamma\nu})V_\beta
\end{aligned}
$$

$$
\begin{aligned}
= {} & R'_E{}^{\beta}{}_{\sigma\nu\mu}V_\beta + R'_{SU(2)}{}^{\beta}{}_{\sigma\nu\mu}V_\beta + R'_{DE}{}^{\beta}{}_{\sigma\nu\mu}V_\beta + R'_{DSU(2)}{}^{\beta}{}_{\sigma\nu\mu}V_\beta + R'_{SU(3)}{}^{\beta}{}_{\sigma\nu\mu}V_\beta + R'_U{}^{\beta}{}_{\sigma\nu\mu}V_\beta + \\
& + R'_V{}^{\beta}{}_{\sigma\nu\mu}V_\beta + R'_S{}^{\beta}{}_{\sigma\nu}V_\beta + R'_\Theta{}^{\beta}{}_{\sigma\nu}V_\beta + R'_B{}^{\beta}{}_{\sigma\nu}V_\beta + R'_G{}^{\beta}{}_{\sigma\nu\mu}V_\beta
\end{aligned}
$$

where all $F_{\ldots}{}^1{}_{\nu\mu}$ and $F_{\ldots}{}^2{}_{\nu\mu}$ terms have summations over the four layers (see below) except the terms $F_S{}^1{}_{\nu\mu} + F_S{}^2{}_{\nu\mu} + F_\Theta{}^1{}_{\nu\mu} + F_\Theta{}^2{}_{\nu\mu} + F_B{}^1{}_{\nu\mu} + F_B{}^2{}_{\nu\mu}$, and where

$$
\begin{aligned}
R'_{SU(3)}{}^{\beta}{}_{\sigma\nu\mu} &= ig^{\beta}{}_{\sigma}(F_{SU(3)}{}^1{}_{\nu\mu} + F_{SU(3)}{}^2{}_{\nu\mu}) \\
R'_{SU(2)}{}^{\beta}{}_{\sigma\nu\mu} &= ig^{\beta}{}_{\sigma}(F_W{}^1{}_{\nu\mu} + F_W{}^2{}_{\nu\mu}) \\
R'_E{}^{\beta}{}_{\sigma\nu\mu} &= ig^{\beta}{}_{\sigma}(F_E{}^1{}_{\nu\mu} + F_E{}^2{}_{\nu\mu}) \\
R'_U{}^{\beta}{}_{\sigma\nu\mu} &= ig^{\beta}{}_{\sigma}(F_U{}^1{}_{\nu\mu} + F_U{}^2{}_{\nu\mu}) \\
R'_V{}^{\beta}{}_{\sigma\nu\mu} &= ig^{\beta}{}_{\sigma}(F_V{}^1{}_{\nu\mu} + F_V{}^2{}_{\nu\mu}) \\
R'_{DSU(2)}{}^{\beta}{}_{\sigma\nu\mu} &= ig^{\beta}{}_{\sigma}(F_{DW}{}^1{}_{\nu\mu} + F_{DW}{}^2{}_{\nu\mu}) \\
R'_{DE}{}^{\beta}{}_{\sigma\nu\mu} &= ig^{\beta}{}_{\sigma}(F_{DE}{}^1{}_{\nu\mu} + F_{DE}{}^2{}_{\nu\mu}) \\
R'_S{}^{\beta}{}_{\sigma\nu\mu} &= ig^{\beta}{}_{\sigma}(F_S{}^1{}_{\nu\mu} + F_S{}^2{}_{\nu\mu}) \\
R'_\Theta{}^{\beta}{}_{\sigma\nu\mu} &= ig^{\beta}{}_{\sigma}(F_\Theta{}^1{}_{\nu\mu} + F_\Theta{}^2{}_{\nu\mu}) \\
R'_B{}^{\beta}{}_{\sigma\nu\mu} &= ig^{\beta}{}_{\sigma}(F_B{}^1{}_{\nu\mu} + F_B{}^2{}_{\nu\mu})
\end{aligned}
$$

and

$$
\begin{aligned}
R'_G{}^{\beta}{}_{\sigma\nu\mu} = {} & \partial_\mu H^{1\beta}{}_{\sigma\nu} - \partial_\nu H^{1\beta}{}_{\sigma\mu} + H^{1\gamma}{}_{\nu\sigma}H^{1\beta}{}_{\gamma\mu} - H^{1\gamma}{}_{\mu\sigma}H^{1\beta}{}_{\gamma\nu} + \partial_\mu H^{2\beta}{}_{\sigma\nu} - \partial_\nu H^{2\beta}{}_{\sigma\mu} + \\
& + H^{2\gamma}{}_{\nu\sigma}H^{2\beta}{}_{\gamma\mu} - H^{2\gamma}{}_{\mu\sigma}H^{2\beta}{}_{\gamma\nu} + H^{1\gamma}{}_{\nu\sigma}H^{2\beta}{}_{\gamma\mu} - H^{1\gamma}{}_{\mu\sigma}H^{2\beta}{}_{\gamma\nu} + H^{2\gamma}{}_{\nu\sigma}H^{1\beta}{}_{\gamma\mu} - \Gamma^{2\gamma}{}_{\mu\sigma}\Gamma^{\beta}{}_{\gamma\nu} \\
= {} & R^{1\beta}{}_{\sigma\nu\mu} + R^{2\beta}{}_{\sigma\nu\mu}
\end{aligned}
$$

with

$$H^{\beta}{}_{\sigma\nu\mu} = \partial_\mu H^{\beta}{}_{\sigma\nu} - \partial_\nu H^{\beta}{}_{\sigma\mu} + H^{\gamma}{}_{\nu\sigma} H^{\beta}{}_{\gamma\mu} - H^{\gamma}{}_{\mu\sigma} H^{\beta}{}_{\gamma\nu}$$
$$R^{1\beta}{}_{\sigma\nu\mu} = \partial_\mu H^{1\beta}{}_{\sigma\nu} - \partial_\nu H^{1\beta}{}_{\sigma\mu} + H^{1\gamma}{}_{\nu\sigma} H^{1\beta}{}_{\gamma\mu} - H^{1\gamma}{}_{\mu\sigma} H^{1\beta}{}_{\gamma\nu}$$
$$R^{2\beta}{}_{\sigma\nu\mu\rho} = \partial_\mu H^{2\beta}{}_{\sigma\nu} - \partial_\nu H^{2\beta}{}_{\sigma\mu} + H^{2\gamma}{}_{\nu\sigma} H^{2\beta}{}_{\gamma\mu} - H^{2\gamma}{}_{\mu\sigma} H^{2\beta}{}_{\gamma\nu} +$$
$$+ H^{1\gamma}{}_{\nu\sigma} H^{2\beta}{}_{\gamma\mu} - H^{1\gamma}{}_{\mu\sigma} H^{2\beta}{}_{\gamma\nu} + H^{2\gamma}{}_{\nu\sigma} H^{1\beta}{}_{\gamma\mu} - H^{2\gamma}{}_{\mu\sigma} H^{1\beta}{}_{\gamma\nu}$$

and

$$H^{1\sigma}{}_{\nu\mu} = \Gamma_{GR}{}^{\sigma}{}_{\nu\mu}$$
$$H^{2\sigma}{}_{\nu\mu} = \Gamma_{GR}{}^{2\sigma}{}_{\nu\mu}$$

and where with summations over four layers indicated by Σ (and layer numbers on fields not shown to avoid clutter) we have

$$F_{SU(3)}{}^{1}{}_{\varkappa\mu} = \Sigma\ \{\partial A_{SU(3)}{}^{1}{}_{\mu}/\partial x^{\varkappa} - \partial A_{SU(3)}{}^{1}{}_{\varkappa}/\partial x^{\mu} + ig_1[A_{SU(3)}{}^{1}{}_{\varkappa}, A_{U(3)}{}^{1}{}_{\mu}]\ \}$$
$$F_{W}{}^{1}{}_{\varkappa\mu} = \Sigma\ \{\partial W^{1}{}_{\mu}/\partial x^{\varkappa} - \partial W^{1}{}_{\varkappa}/\partial x^{\mu} + ig_2[W^{1}{}_{\varkappa}, W^{1}{}_{\mu}]\ \}$$
$$F_{E}{}^{1}{}_{\varkappa\mu} = \Sigma\ \{\partial A_{E}{}^{1}{}_{\mu}/\partial x^{\varkappa} - \partial A_{E}{}^{1}{}_{\varkappa}/\partial x^{\mu}\ \}$$
$$F_{DW}{}^{1}{}_{\varkappa\mu} = \Sigma\ \{\partial W_{D}{}^{1}{}_{\mu}/\partial x^{\varkappa} - \partial W_{D}{}^{1}{}_{\varkappa}/\partial x^{\mu} + ig_4[W_{D}{}^{1}{}_{\varkappa}, W_{D}{}^{1}{}_{\mu}]\ \}$$
$$F_{DE}{}^{1}{}_{\varkappa\mu} = \Sigma\ \{\partial A_{DE}{}^{1}{}_{\mu}/\partial x^{\varkappa} - \partial A_{DE}{}^{1}{}_{\varkappa}/\partial x^{\mu}\}$$
$$F_{U}{}^{1}{}_{\varkappa\mu} = \Sigma\ \{\partial U^{1}{}_{\mu}/\partial x^{\varkappa} - \partial U^{1}{}_{\varkappa}/\partial x^{\mu} + ig_6[U^{1}{}_{\varkappa}, U^{1}{}_{\mu}]\ \}$$
$$F_{V}{}^{1}{}_{\varkappa\mu} = \Sigma\ \{\partial V^{1}{}_{\mu}/\partial x^{\varkappa} - \partial V^{1}{}_{\varkappa}/\partial x^{\mu} + ig_7[V^{1}{}_{\varkappa}, V^{1}{}_{\mu}]\ \}$$
$$F_{S}{}^{1}{}_{\varkappa\mu} = \partial A_{S}{}^{1}{}_{\mu}/\partial x^{\varkappa} - \partial A_{S}{}^{1}{}_{\varkappa}/\partial x^{\mu} + ig_8[A_{S}{}^{1}{}_{\varkappa}, A_{S}{}^{1}{}_{\mu}]$$
$$F_{\Theta}{}^{1}{}_{\varkappa\mu} = \partial A_{\Theta}{}^{1}{}_{\mu}/\partial x^{\varkappa} - \partial A_{\Theta}{}^{1}{}_{\varkappa}/\partial x^{\mu} + ig_{\Theta}[A_{\Theta}{}^{1}{}_{\varkappa}, A_{\Theta}{}^{1}{}_{\mu}]$$
$$F_{B}{}^{1}{}_{\varkappa\mu} = \partial B^{1}{}_{\mu}/\partial x^{\varkappa} - \partial B^{1}{}_{\varkappa}/\partial x^{\mu} + i[B^{1}{}_{\varkappa}, B^{1}{}_{\mu}]$$

$$F_{SU(3)}{}^{2}{}_{\varkappa\mu} = \Sigma\ \{\partial A_{SU(3)}{}^{2}{}_{\mu}/\partial x^{\varkappa} - \partial A_{SU(3)}{}^{2}{}_{\varkappa}/\partial x^{\mu} + ig_1[A_{SU(3)}{}^{2}{}_{\varkappa}, A_{SU(3)}{}^{2}{}_{\mu}] +$$
$$+ ig_1[A_{SU(3)}{}^{1}{}_{\varkappa}, A_{SU(3)}{}^{2}{}_{\mu}] + ig_1[A_{SU(3)}{}^{2}{}_{\varkappa}, A_{SU(3)}{}^{1}{}_{\mu}]\}$$
$$F_{W}{}^{2}{}_{\varkappa\mu} = \Sigma\ \{\partial W^{2}{}_{\mu}/\partial x^{\varkappa} - \partial W^{2}{}_{\varkappa}/\partial x^{\mu} + ig_2[W^{2}{}_{\varkappa}, W^{2}{}_{\mu}] + ig_2[W^{1}{}_{\varkappa}, W^{2}{}_{\mu}] + ig_2[W^{2}{}_{\varkappa}, W^{1}{}_{\mu}]\ \}$$
$$F_{E}{}^{2}{}_{\varkappa\mu} = \Sigma\ \{\partial A_{E}{}^{2}{}_{\mu}/\partial x^{\varkappa} - \partial A_{E}{}^{2}{}_{\varkappa}/\partial x^{\mu}\}$$
$$F_{DW}{}^{2}{}_{\varkappa\mu} = \Sigma\ \{\partial W_{D}{}^{2}{}_{\mu}/\partial x^{\varkappa} - \partial W_{D}{}^{2}{}_{\varkappa}/\partial x^{\mu} + ig_4[W_{D}{}^{2}{}_{\varkappa}, W_{D}{}^{2}{}_{\mu}] + ig_4[W_{D}{}^{1}{}_{\varkappa}, W_{D}{}^{2}{}_{\mu}] +$$
$$+ ig_4[W_{D}{}^{2}{}_{\varkappa}, W_{D}{}^{1}{}_{\mu}]\}$$
$$F_{DE}{}^{2}{}_{\varkappa\mu} = \Sigma\ \{\partial A_{DE}{}^{2}{}_{\mu}/\partial x^{\varkappa} - \partial A_{DE}{}^{2}{}_{\varkappa}/\partial x^{\mu}\}$$
$$F_{U}{}^{2}{}_{\varkappa\mu} = \Sigma\ \{\partial U^{2}{}_{\mu}/\partial x^{\varkappa} - \partial U^{2}{}_{\varkappa}/\partial x^{\mu} + ig_6[U^{2}{}_{\varkappa}, U^{2}{}_{\mu}] + ig_6[U^{1}{}_{\varkappa}, U^{2}{}_{\mu}] + ig_6[U^{2}{}_{\varkappa}, U^{1}{}_{\mu}]\}$$
$$F_{V}{}^{2}{}_{\varkappa\mu} = \Sigma\ \{\partial V^{2}{}_{\mu}/\partial x^{\varkappa} - \partial V^{2}{}_{\varkappa}/\partial x^{\mu} + ig_7[V^{2}{}_{\varkappa}, V^{2}{}_{\mu}] + ig_7[V^{1}{}_{\varkappa}, V^{2}{}_{\mu}] + ig_7[V^{2}{}_{\varkappa}, V^{1}{}_{\mu}]\}$$
$$F_{S}{}^{2}{}_{\kappa\mu} = \partial A_{S}{}^{2}{}_{\mu}/\partial x^{\kappa} - \partial A_{S}{}^{2}{}_{\kappa}/\partial x^{\mu} + ig_8[A_{S}{}^{2}{}_{\kappa}, A_{S}{}^{2}{}_{\mu}] + ig_8[A_{S}{}^{1}{}_{\kappa}, A_{S}{}^{2}{}_{\mu}] +$$

$$+ ig_8[A_S{}^2{}_\kappa, A_S{}^1{}_\mu]$$
$$F_\Theta{}^2{}_{\kappa\mu} = \partial A_\Theta{}^2{}_\mu/\partial x^\kappa - \partial A_\Theta{}^2{}_\kappa/\partial x^\mu + ig_\Theta[A_\Theta{}^2{}_\kappa, A_\Theta{}^2{}_\mu] + ig_\Theta[A_\Theta{}^1{}_\kappa, A_\Theta{}^2{}_\mu] + ig_\Theta[A_\Theta{}^2{}_\kappa, A_\Theta{}^1{}_\mu]$$
$$F_B{}^2{}_{\varkappa\mu} = \partial B^2{}_\mu/\partial x^\varkappa - \partial B^2{}_\varkappa/\partial x^\mu + i[B^2{}_\mu, B^2{}_\varkappa] + i[B^1{}_\mu, B^2{}_\varkappa] + i[B^2{}_\mu, B^1{}_\varkappa]$$

Note that $R'^\beta{}_{\sigma\nu\mu}$ factorizes into $[U(1)\otimes SU(2)\otimes U(1)\otimes SU(2)\otimes SU(3)\otimes U(4)\otimes U(4)]^4\otimes U(4)\otimes U(192)$ parts and a Riemann-Christoffel Gravitational curvature tensor part. For later use in defining a lagrangian we define

$$R'^\beta{}_{\sigma\nu\mu} = R'_E{}^{1\beta}{}_{\sigma\nu\mu} + R'_E{}^{2\beta}{}_{\sigma\nu\mu} + R'_{SU(2)}{}^{1\beta}{}_{\sigma\nu\mu} + R'_{SU(2)}{}^{2\beta}{}_{\sigma\nu\mu} + R'_{DE}{}^{1\beta}{}_{\sigma\nu\mu} + R'_{DE}{}^{2\beta}{}_{\sigma\nu\mu} + R'_{DSU(2)}{}^{1\beta}{}_{\sigma\nu\mu} +$$
$$+ R'_{DSU(2)}{}^{2\beta}{}_{\sigma\nu\mu} + R'_{SU(3)}{}^{1\beta}{}_{\sigma\nu\mu} + R'_{SU(3)}{}^{2\beta}{}_{\sigma\nu\mu} + R'_U{}^{1\beta}{}_{\sigma\nu\mu} + R'_U{}^{2\beta}{}_{\sigma\nu\mu} + R'_V{}^{1\beta}{}_{\sigma\nu\mu} + R'_V{}^{2\beta}{}_{\sigma\nu\mu} +$$
$$+ R'_S{}^{1\beta}{}_{\sigma\nu\mu} + R'_S{}^{2\beta}{}_{\sigma\nu\mu} + R'_\Theta{}^{1\beta}{}_{\sigma\nu\mu} + R'_\Theta{}^{2\beta}{}_{\sigma\nu\mu} + R'_B{}^{1\beta}{}_{\sigma\nu\mu} + R'_B{}^{2\beta}{}_{\sigma\nu\mu} + R^{1\beta}{}_{\sigma\nu\mu} + R^{2\beta}{}_{\sigma\nu\mu}$$

where

$$R'_E{}^{1\beta}{}_{\sigma\nu\mu} = ig^\beta{}_\sigma F_E{}^1{}_{\nu\mu}$$
$$R'_E{}^{2\beta}{}_{\sigma\nu\mu} = ig^\beta{}_\sigma F_{DE}{}^2{}_{\nu\mu}$$

$$R'_{DE}{}^{1\beta}{}_{\sigma\nu\mu} = ig^\beta{}_\sigma F_E{}^1{}_{\nu\mu}$$
$$R'_{DE}{}^{2\beta}{}_{\sigma\nu\mu} = ig^\beta{}_\sigma F_{DE}{}^2{}_{\nu\mu}$$

$$R'_{SU(2)}{}^{1\beta}{}_{\sigma\nu\mu} = ig^\beta{}_\sigma F_W{}^1{}_{\nu\mu}$$
$$R'_{SU(2)}{}^{2\beta}{}_{\sigma\nu\mu} = ig^\beta{}_\sigma F_{DW}{}^2{}_{\nu\mu}$$

$$R'_{DSU(2)}{}^{1\beta}{}_{\sigma\nu\mu} = ig^\beta{}_\sigma F_W{}^1{}_{\nu\mu}$$
$$R'_{DSU(2)}{}^{2\beta}{}_{\sigma\nu\mu} = ig^\beta{}_\sigma F_{DW}{}^2{}_{\nu\mu}$$

$$R'_{SU(3)}{}^{1\beta}{}_{\sigma\nu\mu} = ig^\beta{}_\sigma F_{SU(3)}{}^1{}_{\nu\mu}$$
$$R'_{SU(3)}{}^{2\beta}{}_{\sigma\nu\mu} = ig^\beta{}_\sigma F_{SU(3)}{}^2{}_{\nu\mu}$$

$$R'_U{}^{1\beta}{}_{\sigma\nu\mu} = ig^\beta{}_\sigma F_U{}^1{}_{\nu\mu}$$
$$R'_U{}^{2\beta}{}_{\sigma\nu\mu} = ig^\beta{}_\sigma F_U{}^2{}_{\nu\mu}$$

$$R'_V{}^{1\beta}{}_{\sigma\nu\mu} = ig^\beta{}_\sigma F_V{}^1{}_{\nu\mu}$$
$$R'_V{}^{2\beta}{}_{\sigma\nu\mu} = ig^\beta{}_\sigma F_V{}^2{}_{\nu\mu}$$

$$R'_S{}^{1\beta}{}_{\sigma\nu\mu} = ig^\beta{}_\sigma F_S{}^1{}_{\nu\mu}$$
$$R'_S{}^{2\beta}{}_{\sigma\nu\mu} = ig^\beta{}_\sigma F_S{}^2{}_{\nu\mu}$$

$$R'_{\Theta}{}^{1\beta}{}_{\sigma\nu\mu} = ig^{\beta}{}_{\sigma}F_{\Theta}{}^{1}{}_{\nu\mu}$$
$$R'_{\Theta}{}^{2\beta}{}_{\sigma\nu\mu} = ig^{\beta}{}_{\sigma}F_{\Theta}{}^{2}{}_{\nu\mu}$$

$$R'_{B}{}^{1\beta}{}_{\sigma\nu\mu} = ig^{\beta}{}_{\sigma}B^{1}{}_{\nu\mu}$$
$$R'_{B}{}^{2\beta}{}_{\sigma\nu\mu} = ig^{\beta}{}_{\sigma}B^{2}{}_{\nu\mu}$$

The total Ricci tensor is

$$R'_{\sigma_\mu} = R'^{\beta}{}_{\sigma\beta\mu}$$

$$= iF_{E}{}^{1}{}_{\sigma\mu} + iF_{E}{}^{2}{}_{\sigma\mu} + iF_{W}{}^{1}{}_{\sigma\mu} + iF_{W}{}^{2}{}_{\sigma\mu} + iF_{DE}{}^{1}{}_{\sigma\mu} + iF_{DE}{}^{2}{}_{\sigma\mu} + iF_{DW}{}^{1}{}_{\sigma\mu} + iF_{DW}{}^{2}{}_{\sigma\mu} + iF_{SU(3)}{}^{1}{}_{\sigma\mu} + iF_{SU(3)}{}^{2}{}_{\sigma\mu} +$$
$$+ iF_{U}{}^{1}{}_{\sigma\mu} + iF_{U}{}^{2}{}_{\sigma\mu} + iF_{V}{}^{1}{}_{\sigma\mu} + iF_{V}{}^{2}{}_{\sigma\mu} + iF_{S}{}^{1}{}_{\sigma\mu} + iF_{S}{}^{2}{}_{\sigma\mu} + iF_{\Theta}{}^{1}{}_{\sigma\mu} + iF_{\Theta}{}^{2}{}_{\sigma\mu} + iF_{B}{}^{1}{}_{\sigma\mu} + iF_{B}{}^{2}{}_{\sigma\mu} +$$
$$+ \partial_\mu H^{1\beta}{}_{\sigma\beta} - \partial_\beta H^{1\beta}{}_{\sigma\mu} + H^{1\gamma}{}_{\beta\sigma}H^{1\beta}{}_{\gamma\mu} - H^{1\gamma}{}_{\mu\sigma}H^{1\beta}{}_{\gamma\beta} +$$
$$+ \partial_\mu H^{2\beta}{}_{\sigma\beta} - \partial_\beta H^{2\beta}{}_{\sigma\mu} + H^{2\gamma}{}_{\beta\sigma}H^{2\beta}{}_{\gamma\mu} - H^{2\gamma}{}_{\mu\sigma}H^{2\beta}{}_{\gamma\beta} + H^{1\gamma}{}_{\beta\sigma}H^{2\beta}{}_{\gamma\mu} - H^{1\gamma}{}_{\mu\sigma}H^{2\beta}{}_{\gamma\beta} + H^{2\gamma}{}_{\beta\sigma}H^{1\beta}{}_{\gamma\mu} - H^{2\gamma}{}_{\mu\sigma}H^{1\beta}{}_{\gamma\beta}$$

$$= R'_{E}{}^{1}{}_{\sigma\mu} + R'_{E}{}^{2}{}_{\sigma\mu} + R'_{SU(2)}{}^{1}{}_{\sigma\mu} + R'_{SU(2)}{}^{2}{}_{\sigma\mu} + R'_{DE}{}^{1}{}_{\sigma\mu} + R'_{DE}{}^{2}{}_{\sigma\mu} + R'_{DSU(2)}{}^{1}{}_{\sigma\mu} + R'_{DSU(2)}{}^{2}{}_{\sigma\mu} + R'_{SU(3)}{}^{1}{}_{\sigma\mu} +$$
$$+ R'_{SU(3)}{}^{2}{}_{\sigma\mu} + R'_{U}{}^{1}{}_{\sigma\mu} + R'_{U}{}^{2}{}_{\sigma\mu} + R'_{V}{}^{1}{}_{\sigma\mu} + R'_{V}{}^{2}{}_{\sigma\mu} + R'_{S}{}^{1}{}_{\sigma\mu} + R'_{S}{}^{2}{}_{\sigma\mu} + R'_{\Theta}{}^{1}{}_{\sigma\mu} + R'_{\Theta}{}^{2}{}_{\sigma\mu} + R'_{B}{}^{1\beta}{}_{\sigma\beta\mu} +$$
$$+ R'_{B}{}^{2\beta}{}_{\sigma\beta\mu} + R^{1}{}_{\sigma\mu} + R^{2}{}_{\sigma\mu}$$
$$= R'^{1}{}_{\sigma\mu} + R'^{2}{}_{\sigma\mu}$$

where

$$R'^{1}{}_{\sigma\mu} = R'_{E}{}^{1}{}_{\sigma\mu} + R'_{SU(2)}{}^{1}{}_{\sigma\mu} + R'_{DE}{}^{1}{}_{\sigma\mu} + R'_{DSU(2)}{}^{1}{}_{\sigma\mu} + R'_{SU(3)}{}^{1}{}_{\sigma\mu} + R'_{U}{}^{1}{}_{\sigma\mu} + R'_{V}{}^{1}{}_{\sigma\mu} + R'_{S}{}^{1}{}_{\sigma\mu} +$$
$$+ R'_{\Theta}{}^{1}{}_{\sigma\mu} + R'_{B}{}^{1\beta}{}_{\sigma\beta\mu} + R^{1}{}_{\sigma\mu}$$

$$R'^{2}{}_{\sigma\mu} = R'_{E}{}^{2}{}_{\sigma\mu} + R'_{SU(2)}{}^{2}{}_{\sigma\mu} + R'_{DE}{}^{2}{}_{\sigma\mu} + R'_{DSU(2)}{}^{2}{}_{\sigma\mu} + R'_{SU(3)}{}^{2}{}_{\sigma\mu} + R'_{U}{}^{2}{}_{\sigma\mu} + R'_{V}{}^{2}{}_{\sigma\mu} + R'_{S}{}^{2}{}_{\sigma\mu} +$$
$$+ R'_{\Theta}{}^{2}{}_{\sigma\mu} + R'_{B}{}^{2\beta}{}_{\sigma\beta\mu} + R^{2}{}_{\sigma\mu}$$

with

$$R'_{E}{}^{1}{}_{\sigma\mu} = iF_{E}{}^{1}{}_{\sigma\mu}$$
$$R'_{E}{}^{2}{}_{\sigma\mu} = iF_{E}{}^{2}{}_{\sigma\mu}$$

$$R'_{SU(2)}{}^{1}{}_{\sigma\mu} = iF_{W}{}^{1}{}_{\sigma\mu}$$
$$R'_{SU(2)}{}^{2}{}_{\sigma\mu} = iF_{W}{}^{2}{}_{\sigma\mu}$$

$$R'_{DE}{}^{1}{}_{\sigma\mu} = iF_{DE}{}^{1}{}_{\sigma\mu}$$
$$R'_{DE}{}^{2}{}_{\sigma\mu} = iF_{DE}{}^{2}{}_{\sigma\mu}$$

$$R'_{DSU(2)}{}^{1}{}_{\sigma\mu} = iF_{DW}{}^{1}{}_{\sigma\mu}$$
$$R'_{DSU(2)}{}^{2}{}_{\sigma\mu} = iF_{DW}{}^{2}{}_{\sigma\mu}$$

$$R'_{SU(3)}{}^{1}{}_{\sigma\mu} = iF_{SU(3)}{}^{1}{}_{\sigma\mu}$$
$$R'_{SU(3)}{}^{2}{}_{\sigma\mu} = iF_{SU(3)}{}^{2}{}_{\sigma\mu}$$

$$R'_{U}{}^{1}{}_{\sigma\mu} = iF_{U}{}^{1}{}_{\sigma\mu}$$
$$R'_{U}{}^{2}{}_{\sigma\mu} = iF_{U}{}^{2}{}_{\sigma\mu}$$

$$R'_{V}{}^{1}{}_{\sigma\mu} = iF_{V}{}^{1}{}_{\sigma\mu}$$
$$R'_{V}{}^{2}{}_{\sigma\mu} = iF_{V}{}^{2}{}_{\sigma\mu}$$

$$R'_{S}{}^{1}{}_{\sigma\mu} = iF_{S}{}^{1}{}_{\sigma\mu}$$
$$R'_{S}{}^{2}{}_{\sigma\mu} = iF_{S}{}^{2}{}_{\sigma\mu}$$
$$R'_{\Theta}{}^{1}{}_{\sigma\mu} = iF_{\Theta}{}^{1}{}_{\sigma\mu}$$
$$R'_{\Theta}{}^{2}{}_{\sigma\mu} = iF_{\Theta}{}^{2}{}_{\sigma\mu}$$

$$R'_{B}{}^{1}{}_{\sigma\mu} = iF_{B}{}^{1}{}_{\sigma\mu}$$
$$R'_{B}{}^{2}{}_{\sigma\mu} = iF_{B}{}^{2}{}_{\sigma\mu}$$

with the further definition of $R''^{1}{}_{\sigma\mu}$ and $R''^{2}{}_{\sigma\mu}$:

$$R''^{1}{}_{\sigma\mu} = R'_{SU(3)}{}^{1}{}_{\sigma\mu} + R^{1}{}_{\sigma\mu}$$
$$R''^{2}{}_{\sigma\mu} = R'_{SU(3)}{}^{2}{}_{\sigma\mu} + R^{2}{}_{\sigma\mu}$$

$R'^{1}{}_{\sigma\mu}$ is the Ricci tensor. An additional Ricci-like tensor is

$$H_{\sigma_\mu} = H^{\beta}{}_{\sigma\beta\mu}$$

The curvature scalar is

$$\begin{aligned} R' = g^{\sigma\mu}R'_{\sigma\mu} = & + \partial^{\sigma}H^{1\beta}{}_{\sigma\beta} - \partial_{\beta}H^{1\beta}{}_{\sigma}{}^{\sigma} + H^{1\gamma}{}_{\beta\sigma}H^{1\beta}{}_{\gamma}{}^{\sigma} - H^{1\gamma}{}_{\mu\sigma}H^{1\beta}{}_{\gamma\beta} + \partial^{\sigma}H^{2\beta}{}_{\sigma\beta} - \partial_{\beta}H^{2\beta}{}_{\sigma}{}^{\sigma} + \\ & + H^{2\gamma}{}_{\beta\sigma}H^{2\beta}{}_{\gamma}{}^{\sigma} - H^{2\gamma\sigma}{}_{\sigma}H^{2\beta}{}_{\gamma\beta} + H^{1\gamma}{}_{\beta\sigma}H^{2\beta}{}_{\gamma}{}^{\sigma} - H^{1\gamma\sigma}{}_{\sigma}H^{2\beta}{}_{\gamma\beta} + H^{2\gamma}{}_{\beta\sigma}H^{1\beta}{}_{\gamma}{}^{\sigma} - H^{2\gamma\sigma}{}_{\sigma}H^{1\beta}{}_{\gamma\beta} \end{aligned}$$

$$= g^{\sigma\mu}(R^{1\beta}{}_{\sigma\beta\mu} + R^{2\beta}{}_{\sigma\beta\mu})$$

4.2.16.2 Vector Boson and Graviton Lagrangian Terms

The vector boson and gravitational part of the lagrangian of the Unified SuperStandard (with the Higgs sector and the Faddeev-Popov terms gauge sector not displayed here) is:

$$\mathcal{L} = \mathrm{Tr}\ \sqrt{g}[M D_\nu R''^{1}{}_{\sigma\mu} D^\nu R''^{2\sigma\mu} + aR'^{1}{}_{\sigma\mu}R'^{2\sigma\mu} + bR' + cg^{\sigma\mu}g^{2}{}_{\sigma\mu} + c'g^{2\sigma\mu}g^{2}{}_{\sigma\mu} - dA_{SU(3)}{}^{2}{}_{\mu}A_{SU(3)}{}^{2\mu}]$$

where M, a, b, c, c', and d are constants, and $R''^{i}{}_{\sigma\mu}$ for i = 1, 2 determined above.[125]

This higher derivative lagrangian maintains the locality of the theory but does entail a modest modification in the derivation of the Euler-Lagrange equations of motion. It also requires the use of principal value propagators rather than ordinary Feynman propagators for gluon and graviton interactions. Thus the Strong Interaction sector, and the Gravitation sector are Action-at-a-Distance theories that are similar in spirit to Wheeler-Feynman Electrodynamics. The two U(1) Electromagnetic sectors, the Generation group U(4) gauge field sector, the Layer group U(4) gauge field sector, the two SU2) Weak sectors, the U(4) A_S gauge field sector, the spinor connection sector, and the Θ-interaction sector may, or may not, be Action-at-a-Distance fields. They are not constrained to be Action-at-a-Distance by the present considerations.

Since we wish to apply our theory cosmologically, and within hadrons, where the gravitational spinor connections are negligible due to the smallness of the gravitational constant G and the 'smallness' of B spin on the cosmological scale, we set $F^{1}{}_{\nu\mu} = F^{2}{}_{\nu\mu} = 0$ and find[126]

---

[125] One may ask why $R''^{1}{}_{\sigma\mu}$ and $R''^{2}{}_{\sigma\mu}$ appear in the first term of the lagrangian, and not other interaction terms. We believe the primary reason is: "The extended vierbein $1^{\mu ai}(x)$ can be viewed as located at a point in a higher dimensional complex-valued space.

$$1^{\mu ai}(x) = (\partial\xi_X{}^{ai}(x)/\partial x_\mu)_{X=h(x)}$$

where $\xi_X{}^{ai}$ is a set of locally inertial coordinates located at point X, and x = h(x) is a 4-dimensional point in a tangent subspace of the higher dimensional space:

$$X = h(x)$$

The relation between complex 4-dimensional coordinates x and the higher dimensional coordinates X is an embedding of a 4-dimensional surface within the higher dimensional complex space when account is taken of the range of possible x values. We have considered such embeddings in Blaha (2015a), and in earlier books, and developed a theory of a higher dimensional complex-valued space (the *Megaverse*) that contains our universe and probably many other universes." Thus SU(3) and Gravitation have a special role in our particle dynamics based on geometry. The second reason is the common feature of color SU(3) and real-valued General Relativity is that they are the only interactions that do not participate in 'rotations of interactions' as described earlier and in chapter 31 of Blaha (2017b). The third, practical reason is the experimental reality that the Strong Interaction and Gravitation are known to have 'anomalous' features that will be seen to be remedied by these insertions while the other interactions are 'conventional.'

[126] The constants have the dimensions: M has the dimension of inverse mass squared, b has dimension mass squared, a is dimensionless, c and c' have dimension mass, and d has dimension mass squared.

$$\mathcal{L} = \mathrm{Tr}\ \sqrt{g}[MD_\nu(R'^1{}_{SU(3)\sigma\mu} + R'_G{}^1{}_{\sigma\mu})D^\nu(R'_{SU(3)}{}^{2\sigma\mu} + R'_G{}^{2\sigma\mu}) + \\ + aR'^1{}_{\sigma\mu}R'^{2\sigma\mu} + bR' + cg^{\sigma\mu}g^2{}_{\sigma\mu} + c'g^{2\sigma\mu}g^2{}_{\sigma\mu} - dA_{SU(3)}{}^2{}_\mu A_{SU(3)}{}^{2\mu}]$$

Since there are no strong interaction fields in 'empty' space and gravity is negligible within hadrons,[127] we can drop the interaction terms between the Strong interaction and the Gravity interaction. However, we cannot drop the interaction terms amongst Electromagnetism, the Weak interaction, the Strong Interaction, the Generation group U(4) interaction, the U(4) Layer groups interactions, the U(4) Species group interaction, and the U(192) Θ-interaction – within, and between, hadrons. The interaction terms between Electromagnetism and Gravitation are important cosmologically.

The above lagrangian terms can therefore be expressed as:[128]

$$\mathcal{L} = \mathcal{L}_E + \mathcal{L}_{SU(2)} + \mathcal{L}_{DE} + \mathcal{L}_{DSU(2)} + \mathcal{L}_{SU(3)} + \mathcal{L}_U + \mathcal{L}_V + \mathcal{L}_S + \mathcal{L}_\Theta + \mathcal{L}_G + \mathcal{L}_{int}$$

where taking traces of $\mathcal{L}$s terms is understood

$$\mathcal{L}_E = \mathrm{Tr}\ \sqrt{g}\{M\{[\partial_\nu + i(A_E{}^1{}_\nu + A_E{}^2{}_\nu)]F^1{}_{E\sigma\mu}[\partial^\nu + i(A_E{}^{1\nu} + A_E{}^{2\nu})]F^2{}_E{}^{\sigma\mu}\} + aF_E{}^1{}_{\sigma\mu}F_E{}^{2\sigma\mu}\}$$

$$\mathcal{L}_{SU(2)} = \mathrm{Tr}\ \sqrt{g}[aF_W{}^1{}_{\sigma\mu}F_W{}^{2\sigma\mu}]$$
$$\mathcal{L}_{DE} = \mathrm{Tr}\ \sqrt{g}\{M\{[\partial_\nu + i(A_{DE}{}^1{}_\nu + A_{DE}{}^2{}_\nu)]F^1{}_{DE\sigma\mu}[\partial^\nu + i(A_{DE}{}^{1\nu} + A_{DE}{}^{2\nu})]F_{DE}{}^{2\sigma\mu}\} + aF_{DE}{}^1{}_{\sigma\mu}F_{DE}{}^{2\sigma\mu}\}$$
$$\mathcal{L}_{DSU(2)} = \mathrm{Tr}\ \sqrt{g}[aF_W{}^1{}_{\sigma\mu}F_W{}^{2\sigma\mu}]$$
$$\mathcal{L}_{SU(3)} = \mathrm{Tr}\ \sqrt{g}\{M[\partial_\nu + i(A_{SU(3)}{}^1{}_\nu + A_{SU(3)}{}^2{}_\nu)]F_{SU(3)}{}^1{}_{\sigma\mu}[\partial^\nu + i(A_{SU(3)}{}^{1\nu} + A_{SU(3)}{}^{2\nu})]F_{SU(3)}{}^{2\sigma\mu} + \\ + aF_{SU(3)}{}^1{}_{\sigma\mu}F_{SU(3)}{}^{2\sigma\mu} - dA_{SU(3)}{}^2{}_\mu A_{SU(3)}{}^{2\mu}\}$$
$$\mathcal{L}_U = \mathrm{Tr}\ \sqrt{g}[aF_U{}^1{}_{\sigma\mu}F_U{}^{2\sigma\mu}]$$
$$\mathcal{L}_V = \mathrm{Tr}\ \sqrt{g}[aF_V{}^1{}_{\sigma\mu}F_V{}^{2\sigma\mu}]$$
$$\mathcal{L}_S = \mathrm{Tr}\ \sqrt{g}[aF_S{}^1{}_{\sigma\mu}F_S{}^{2\sigma\mu}]$$
$$\mathcal{L}_\Theta = \mathrm{Tr}\ \sqrt{g}[aF_\Theta{}^1{}_{\sigma\mu}F_\Theta{}^{2\sigma\mu}]$$
$$\mathcal{L}_G = \mathrm{Tr}\ \sqrt{g}[MD_\nu R^1{}_{\sigma\mu}D^\nu R^{2\sigma\mu} + aR^1{}_{\sigma\mu}R^{2\sigma\mu} + bg^{\sigma\mu}(R^{1\beta}{}_{\sigma\beta\mu} + R^{2\beta}{}_{\sigma\beta\mu}) + cg^{\sigma\mu}g^2{}_{\sigma\mu} + c'g^{2\sigma\mu}g^2{}_{\sigma\mu}] \\ = \mathrm{Tr}\ \sqrt{g}[MD_\nu R^1{}_{\sigma\mu}D^\nu R^{2\sigma\mu} + aR^1{}_{\sigma\mu}R^{2\sigma\mu} + bH + cg^{\sigma\mu}g^2{}_{\sigma\mu} + c'g^{2\sigma\mu}g^2{}_{\sigma\mu}]$$

$$\mathcal{L}_{int} = \mathcal{L} - (\mathcal{L}_E + \mathcal{L}_{SU(2)} + \mathcal{L}_{DE} + \mathcal{L}_{DSU(2)} + \mathcal{L}_{SU(3)} + \mathcal{L}_U + \mathcal{L}_V + \mathcal{L}_S + \mathcal{L}_\Theta + \mathcal{L}_G)$$

with appropriate sums over layers. Thus $\mathcal{L}_{SU(3)}$, $\mathcal{L}_{SU(2)}$, $\mathcal{L}_E$, $\mathcal{L}_{DE}$, $\mathcal{L}_{DSU(2)}$, $\mathcal{L}_U$,, $\mathcal{L}_V$, $\mathcal{L}_S$, $\mathcal{L}_\Theta$, and parts of $\mathcal{L}_{int}$ are the dominant interactions within hadrons, and $\mathcal{L}_G$, $\mathcal{L}_E$ and parts of $\mathcal{L}_{int}$ are the dominant interactions in space within the framework of this discussion.

---

[127] We show gravity weakens at very short distances using our Two-Tier Quantum Field Theory formalism. See Appendix A, and Blaha (2003) and (2005a) among other books by the author.
[128] We only consider the gauge field lagrangian terms.

The $D_\nu R^1{}_{\sigma\mu}$ and $D^\nu R^{2\sigma\mu}$ terms have the form:

$$D_\nu R^i{}_{\sigma\mu} = + \partial_\nu R^i{}_{\sigma\mu} \;\; - H^{1\beta}{}_{\sigma\nu} R^i{}_{\beta\mu} - H^{1\beta}{}_{\nu\mu} R^i{}_{\sigma\beta}$$

for i = 1, 2.

4.2.16.3 New Vector Boson Interactions

The above lagrangian can be broken up into pieces in the following manner:

$$\mathcal{L}_E = \text{Tr}\,\sqrt{g}\{M\{[\partial_\nu + i(A_E{}^1{}_\nu + A_E{}^2{}_\nu)]F^1{}_{E\sigma\mu}[\partial^\nu + i(A_E{}^{1\nu} + A_E{}^{2\nu})]F^2{}_E{}^{\sigma\mu}\} + aF_E{}^1{}_{\sigma\mu}F_E{}^{2\sigma\mu}\}$$
$$\mathcal{L}_{SU(2)} = \text{Tr}\,\sqrt{g}[aF_W{}^1{}_{\sigma\mu}F_W{}^{2\sigma\mu}]$$
$$\mathcal{L}_{DE} = \text{Tr}\,\sqrt{g}\{M\{[\partial_\nu + i(A_{DE}{}^1{}_\nu + A_{DE}{}^2{}_\nu)]F^1{}_{DE\sigma\mu}[\partial^\nu + i(A_{DE}{}^{1\nu} + A_{DE}{}^{2\nu})]F_{DE}{}^{2\sigma\mu}\} + aF_{DE}{}^1{}_{\sigma\mu}F_{DE}{}^{2\sigma\mu}\}$$
$$\mathcal{L}_{DSU(2)} = \text{Tr}\,\sqrt{g}[aF_W{}^1{}_{\sigma\mu}F_W{}^{2\sigma\mu}]$$
$$\mathcal{L}_{SU(3)} = \text{Tr}\,\sqrt{g}\{M[\partial_\nu + i(A_{SU(3)}{}^1{}_\nu + A_{SU(3)}{}^2{}_\nu)]F_{SU(3)}{}^1{}_{\sigma\mu}[\partial^\nu + i(A_{SU(3)}{}^{1\nu} + A_{SU(3)}{}^{2\nu})]F_{SU(3)}{}^{2\sigma\mu} + $$
$$+ aF_{SU(3)}{}^1{}_{\sigma\mu}F_{SU(3)}{}^{2\sigma\mu} - dA_{SU(3)}{}^2{}_\mu A_{SU(3)}{}^{2\mu}\}$$
$$\mathcal{L}_U = \text{Tr}\,\sqrt{g}[aF_U{}^1{}_{\sigma\mu}F_U{}^{2\sigma\mu}]$$
$$\mathcal{L}_V = \text{Tr}\,\sqrt{g}[aF_V{}^1{}_{\sigma\mu}F_V{}^{2\sigma\mu}]$$
$$\mathcal{L}_S = \text{Tr}\,\sqrt{g}[aF_S{}^1{}_{\sigma\mu}F_S{}^{2\sigma\mu}]$$
$$\mathcal{L}_\Theta = \text{Tr}\,\sqrt{g}[aF_\Theta{}^1{}_{\sigma\mu}F_\Theta{}^{2\sigma\mu}]$$
$$\mathcal{L}_G = \text{Tr}\,\sqrt{g}[MD_\nu R^1{}_{\sigma\mu}D^\nu R^{2\sigma\mu} + aR^1{}_{\sigma\mu}R^{2\sigma\mu} + bg^{\sigma\mu}(R^{1\beta}{}_{\sigma\beta\mu} + R^{2\beta}{}_{\sigma\beta\mu}) + cg^{\sigma\mu}g^2{}_{\sigma\mu} + c'g^{2\sigma\mu}g^2{}_{\sigma\mu}]$$
$$= \text{Tr}\,\sqrt{g}[MD_\nu R^1{}_{\sigma\mu}D^\nu R^{2\sigma\mu} + aR^1{}_{\sigma\mu}R^{2\sigma\mu} + bH + cg^{\sigma\mu}g^2{}_{\sigma\mu} + c'g^{2\sigma\mu}g^2{}_{\sigma\mu}]$$

$$\mathcal{L}_{int} = \mathcal{L} - (\mathcal{L}_E + \mathcal{L}_{SU(2)} + \mathcal{L}_{DE} + \mathcal{L}_{DSU(2)} + \mathcal{L}_{SU(3)} + \mathcal{L}_U + \mathcal{L}_V + \mathcal{L}_S + \mathcal{L}_\Theta + \mathcal{L}_G)$$

again with appropriate sums over layers. Thus $\mathcal{L}_{SU(3)}$, $\mathcal{L}_{SU(2)}$, $\mathcal{L}_E$, $\mathcal{L}_{DE}$, $\mathcal{L}_{DSU(2)}$, $\mathcal{L}_U$,, $\mathcal{L}_V$, $\mathcal{L}_S$, $\mathcal{L}_\Theta$, and parts of $\mathcal{L}_{int}$ are the dominant interactions within hadrons, and $\mathcal{L}_G$, $\mathcal{L}_E$ and parts of $\mathcal{L}_{int}$ are the dominant interactions in space within the framework of this discussion. The terms of $\mathcal{L}_{int}$ have 'new' interactions between gauge fields that are described in some detail in Blaha (2017b). These interactions are not in the conventional Standard Model. They lead to modifications of gravity, the Strong Interactions, spin dynamics and so on.

4.2.17 SuperSymmetry

The Unified SuperStandard Model has SuperSymmetric aspects embedded within it. It has equal numbers of fermions and vector bosons -192 of each.[129] The vector bosons[130] are directly related to the form of the fermion particles' structure – the Fermion Periodic Table. The 192 fermions form a 192 representation of the Θ-Symmetry group, and the 192 vector bosons form another 192 representation of the U(192) Θ-Symmetry group.

4.2.17.1 The Θ-Symmetry Group

The U(192) Θ-Symmetry group has 192^2 gauge fields and generators. Elements of the group algebra form a closed set under commutation. The fundamental representation has 192 dimensions.

4.2.17.2 Fermionic Operators

One can define two sets[131] of 192^2 fermionic operators $\{\mathbf{A}_{\Theta F}{}^{\mu}{}_{nm}(x)\}$ and $\{\mathbf{A}_{\Theta B}{}^{\mu}{}_{nm}(x)\}$ that transform fermions into bosons between the 192 Θ-Symmetry fermion and boson representations; and that also transform bosons into fermions between the 192 Θ-Symmetry fermion and boson representations respectively:

$$\mathbf{A}_{\Theta F}{}^{\mu}{}_{nm}(x)\psi_m(x) \rightarrow A^{\mu}{}_n$$
$$\mathbf{A}_{\Theta B}{}^{\mu}{}_{nm}(x)A_{\mu m}(x) \rightarrow \psi_n$$

where n = 1, ... , 192 and m = 1, ... , 192 label multiplet fields irrespective of internal symmetries. Each n and m pair label one of 192^2 fermionic operators.

The fermionic operators are closed under anticommutation. One representation of these operators in terms of the fermionic and bosonic fields is

$$\mathbf{A}_{\Theta F}{}^{\mu}{}_{nm}(x) = \mathbf{A}^{\mu}{}_n(x)\psi_m(x)$$
$$\mathbf{A}_{\Theta B}{}^{\mu}{}_{nm}(x) = \psi_n(x)\mathbf{A}^{\mu}{}_m(x)^{\dagger}$$

where n and m are labels ranging from 1, ... 192 for fermions and bosons, $\mathbf{A}^{\mu}{}_n(x)$ is the nth of the 192 vector fields, $\psi_m(x)$ is the mth of the 192 fermion fields in eq. 3.2; and $\mathbf{A}^{\mu}{}_m(x)^{\dagger}$ is the hermitean conjugate of the mth vector field of the 192 vector fields and $\psi_n(x)$ is the nth fermion

[129] The 16 Species group vector bosons 'do not count' because they follow from Complex General Relativity.
[130] There are also Higgs bosons and Θ-Symmetry vector bosons. These vector bosons do not affect the number or form of the Fermion Periodic Table and thus can be viewed as a separate issue. In fact we will show the Θ-Symmetry group plays a role in establishing a SuperSymmetry-like formalism in the Unified SuperStandard Model.
[131] While these operators form a Jordan algebra they do not have group representations.

field of the 192 fermion fields. Spatial integrals of the above operators have simpler anticommutation relations.

Elements of the Θ-Symmetry group can be used to map the fermionic operators to different forms. For example if U_Θ is an element of the Θ-Symmetry group we can redefine fermion operators as a different map between fermions and bosons with expresions like

$$\mathbf{A'}_{\Theta F}{}^{\mu}{}_{nm} = U_\Theta \mathbf{A}_{\Theta F}{}^{\mu}{}_{nm} U_\Theta{}^{-1}$$
$$\mathbf{A'}_{\Theta B}{}^{\mu}{}_{nm} = U_\Theta \mathbf{A}_{\Theta B}{}^{\mu}{}_{nm} U_\Theta{}^{-1}$$

Thus one can SuperSymmetric map between the fermion and vector boson multiplets. We have not displayed internal symmetry indices on the fields. Clearly the difference between the internal symmetry fundamental representation of the fermions and the adjoint representation of the vector bosons makes the fermionic transformations 'sluff' over internal symmetries.

4.2.17.3 Relation of the Θ-Symmetry Fields and the Fermionic Fields

The fermionic field operators that we have defined can be related to the U(192) Θ-Symmetry field operators which we will denote $\mathbf{A}_\Theta{}^{\mu}{}_{nm}(x)$ where n = 1, ... , 192 and m = 1, ... , 192 are labels on the <u>192</u> fermion or boson fields. The following relations hold:

1. The product of $A_{\Theta F}{}^{\mu}{}_{nm}(x)$ and $A_{\Theta F}{}^{\mu}{}_{mp}(x)^\dagger$ can be expressed as an equivalent sum of Θ-Symmetry field operators.

2. The product of $A_{\Theta B}{}^{\mu}{}_{nm}(x)^\dagger$ and $A_{\Theta B}{}^{\mu}{}_{mp}(x)$ can be expressed as an equivalent sum of Θ-Symmetry field operators.

3. The product of $A_{\Theta F}{}^{\mu}{}_{nm}(x)$ and $A_{\Theta B}{}^{\mu}{}_{mp}(x)$ can be expressed as an equivalent sum of Θ-Symmetry field operators.

Thus we find that the U(192) Θ-Symmetry field operators can be obtained from products of the fermionic operators.

We also note that the product of a fermionic operator and a Θ-Symmetry field operator is a fermionic operator.

These relations establish a close connection between fermionic field operators and the Θ-Symmetry field operators.

We conclude that there is a SuperSymmetry relation between the fermion and vector boson multiplets in the Unified SuperStandard Model. The benefits of SuperSymmetric transformations in this context remain to be determined. However there is a hint of a significant role in the Megaverse.

4.2.18 Embedded Strings

The Unified SuperStandard Model uses Two-Tier Quantum Field Theory to embed strings in each fermion and boson. See chapter 2 for a discussion of embedding and its String analogue.

4.2.19 Status of the Derivation at this Point

The derivation that we have developed has generated the known form of the Standard Model and Gravitation in our universe. It has added Dark matter and energy sectors to create The Unified SuperStandard Model based on the appearance of conserved and partially conserved particle number operators such as Baryon Number, Layer Number operators and so on. We have introduced the 'Rotation of Imteractions' Θ-symmetry group and showed that ElectroWeak Theory with the Weinberg angle is a simple application of this concept. We have also shown that Θ-symmetry is closely related to a form of SuperSymmetry that rotates between fermions and vector bosons without the introduction of unseen supplementary multiplets of s=particles such as squarks.

The Two-Tier formalism that is at the heart of making the Unified SuperStandard Model finite to all orders in perturbation theory has been shown to be a form of SuperString theory with strings embedded within each fermion and boson.

We now turn to consider the Megaverse – a higher dimensional space containing our universe and (countless?) other universes.

4.2.20 The Megaverse

The Megaverse is a higher dimensional space that contains our universe and other universes as surfaces within it. In this subsection we will describe aspects of its features which were discussed in detail in Blaha (2017c) and described *features of our universe that suggested that it had contact with or close proximity to other universes.*

4..2.20.1 Dimension of the Megaverse

In chapter 1 of this book we showed there was good reason to use the number of gauge fields of particle interactions as the dimension of a universe. We therefore made this an axiom of our theory. The Unified SuperStandard Model has 192 generators in the particle interaction sector with symmetry $[SU(3)\otimes SU(2)\otimes U(1)\otimes SU(2)\otimes U(1)\otimes U(4)\otimes U(4)]^4$. The other fields for Θ-symmetry and Complex General Relativity (which includes the Species group) were not counted in the total by assumption since they were secondary to the primary 192 gauge fields. Their fields did not influence the form of the Fermion Periodic Table – a fact that we consider an indication of their secondary nature.

With this axiom in mind and the symmetry group above, we choose the Megaverse dimension D = 192.[132]

4.2.20.2 Universes as Mass-Energy Islands

In developing the theory of the Megaverse we view universes as islands of mass-energy that maintain their 'integrity' as surfaces due to the gravitational force. Gravity holds universes together rather like molecular forces within a water droplet hold the water molecules together. This effective force gives droplets cohesion as they (perhaps) descend through the earth's atmosphere. It is the origin of *surface tension* in water.

Similarly gravity holds the higher density mass-energy universes together and gives rise to gravity surface tension. In a model presented in Blaha (2017c) for the Big Bang and the expansion of the universe within the Megaverse we found the mass-energy density of the universe was a factor of 10^{30} more than the density in the surrounding Megaverse space.

Thus we have good reason to study the surface tension of universes.

4.2.20.3 Confinement of universes due to 'Surface Tension'

We will assume that the other universes have the same physics as our universe with the possible difference that they may have differing coupling constants and particle masses. As we will discuss later, every point of a universe in a higher dimensional Megaverse has Megaverse points in any neighborhood of the point (except for neighborhoods strictly within the universe).

[132] We use the symbol D to represent the dimension in Blaha (2017c).

Thus we confront the question: what keeps mass-energy at points in a universe or is there leakage from the universe into the Megaverse?

If there is no leakage into the Megaverse then, since each point in the universe is part of a Megaverse surface, one can only assume that there is a barrier to movement into the Megaverse. Taking a note from fluid dynamics, and viewing Megaverse space as one 'material' and the universe as a different 'material,'[133] we can view the barrier as due to 'surface tension.'[134] The Megaverse "exerts a force" confining the contents of the universe to within itself.[135]

The surface tension[136] of a universe γ satisfies the relation

$$\gamma = W/\Delta A$$

where γ is expresased in erg/cm^2, W is the Work, and ΔA is the Area upon which the work is exerted. The pressure Δp exerted by the surface tension for a 'spherical' surface area is

$$\Delta p = 2\gamma/R$$

where R equals the radius of curvature of the surface. The above equations embody the concept that the surface tension force equals the pressure difference at the surface.

If the universe is flat then the surface pressure approaches ∞ giving confinement of fields and particles to the universe.

$$R \rightarrow 0 \quad \text{implies } \Delta p \rightarrow \infty$$

Thus we have the theorem:

Theorem: A universe has no leakage of fields or particles if it is exactly flat.

This theorem is particularly interesting in the case of our universe. It appears to be flat (or very close to flat). The flatness of our universe may be the reason no leakage of fields or particles from our universe has been detected to high accuracy.

If a universe is found with a non-zero radius of curvature then one can expect that some fields and particles may emerge from it into the Megaverse.

While a zero radius of curvature prevents the exit of fields and particles from a universe, it does not prevent the entry of mass-energy into the universe from the Megaverse.

---

[133] Meaning material with much higher mass-energy density and consequently larger internal gravitational attraction.

[134] See Landau (1987).

[135] Although in actuality it is the universe that holds itself together by gravitation.

[136] A useful analogy: the Megaverse is a pool of water; a universe is a denser oil bubble within it. Surface tension caused by the cohesiveness of the oil molecules in the bubble makes it spherical (confines it to a spherical shape). Similarly a universe (denser than the Megaverse) is 'confined' within the Megaverse.

Thus our Continuous creation model of chapter 14 in Blaha (2017c) may be relevant. Entry is possible; exit is forbidden in this case.

4.2.20.4 Quantum Fields 'Emanating' from a Universe

If the curvature of the universe is zero (open universe) then no fields emanate from it. If the curvature of the universe is non-zero (closed universe) then fields may 'leak' into the Megaverse. Then continuity conditions between a universe field and its Megaverse equivalent becomes of interest.

4.2.20.5 Megaverse Dynamics due to Universe Fields

The particles and fields that we have found in our universe will also exist in the Megaverse but the form of their fourier expansions will be 192-dimensional. There will also be universe particles – quantum fields for entire universes.

The universe particles within a Megaverse have dynamical motions in the Megaverse due to the forces exerted between them as well as Megaverse gravity.

In Blaha (2017c) and earlier books we described features of the Wheeler-DeWitt equation that suggested that universes could be viewed as particles or anti-particles, or tachyons. The solutions of this equation are scalar wave functions on a manifold that are analogous to the solutions of the Klein-Gordon equation. The issues of negative probabilities, possible tachyonic solutions, and negative frequency solutions suggest a need for an appropriate particle interpretation of universes that can possibly resolve these problems.

Some physicists have taken the Wheeler-DeWitt equation as the starting point for a theory of a universe as a particle. The Wheeler-DeWitt equation describes the interior of a universe in a quantum framework.

We will take a different approach using the Megaverse as the environment of universe particles that internally have Quantum Gravity, and externally have Megaverse Quantum Gravity.

We view a universe as an extended particle and begin by ignoring the detailed inner structure of universes. This approach is similar to the historical treatment of hadrons such as the proton as particles and developing a theory of them as fundamental particles using form factors, structure functions and so on to approximate their inner structure. Afterwards, as detailed data became available, the detailed investigation of the internal structure of hadrons using quark-parton models followed. We will pursue a similar theoretical development beginning with a theory of universes as extended particles in the 192-dimension Megaverse. The internal structure of the particle universes will eventually be specified by the Wheeler-DeWitt equation expressed in Megaverse coordinates.

The two simplest choices for the nature of universes are "spin 0" *bosonic universes* and *fermionic universes* with odd half integer spin, s_M.[137] We will first consider the possibility of fermionic universes, and then briefly consider "spin 0" *bosonic* universes.

The first issue of fermionic universes (reminiscent of the discussions of spin in the 1920's) is the interpretation of spin states. We suggest that the upper $2^{D/2-1}$ components (with $2^{D/2-2}$ "spin up" and $2^{D/2-2}$ "spin down" states) of a fermionic universe wave function represent a left-handed universe with an excess number of baryons.[138] The lower ($2^{D/2-1}$) components lead to right-handed anti-universes where there is an excess of anti-baryons. These associations are analogous to the interpretations of the Dirac electron wave function.[139]

The universe particle "spin up" and "spin down" states are distinguished by their interactions with gauge fields in a manner analogous to quantum electrodynamics.

4.2.20.6 "Free Field" Dynamics of Fermionic Universe Particles

We now consider universes as extended particles with an odd half integer spin – *fermionic universe particles* - in the D-dimensional Megaverse. In the Megaverse there are D 'Dirac' matrices with $2^{D/2}$ rows and $2^{D/2}$ columns that are the equivalent of the four Dirac matrices in four dimensions. We will denote these D matrices as $\gamma_M{}^i$ for i = 1, 2, … , D. They satisfy the anti-commutation relations:

$$\{\gamma_M{}^i, \gamma_M{}^j\} = 2\,\delta^{ij}$$

and thus form a Clifford algebra. We will choose y^D to be the time coordinate and thus make it pure imaginary with a Reality group transformation. (The D-dimensional Megaverse space is a complex Euclidean space.) Therefore γ^D will be hermitean ($(\gamma^D)^2 = 1$), and the γ^i matrices for i = 1, … , (D – 1) will be anti-hermitean with $(\gamma^i)^2 = -1$. The number of linearly independent matrices in D dimensions is 2^D.

The Megaverse metric is (by use of the Reality group) chosen to be

$$g^{ij} = -\delta^{ij}, \qquad g^{D,D} = 1$$

for i, j = 1, 2, … , (D – 1); and zero otherwise.

Except for the additional dimensions, fermion dynamics is quite similar to the 4-dimensional case. The free universe particle Dirac equation is

---

137 Since the Megaverse is D-dimensional, the spin of fermionic universe particles was shown to be s_M in Blaha (2017c) in chapter 3.
138 D = 192.
139 It is known that phenomena in our universe tend to be left-handed. If this feature of our universe's phenomena is also a property of the universe itself, then, since handedness is an attribute of spin, the treatment of a universe as having spin is not unreasonable.

$$(i\gamma^i\partial_i + m)\psi(y) = 0$$

summed over i = 1, 2, … , D where the mass is assumed to be constant, and set below. The derivative operator, is based on the use of quantum coordinates[140]

$$Y^i(y) = y^i + i\, Y_u{}^i(y)/M_u{}^{D/2}$$

for i = 1, … , D and is defined to be

$$\partial_i = \partial/\partial Y^i(y) = \partial/\partial(y_i - Y_{ui}(y)/M_u{}^{D/2})$$

where *we assume $M_u = M_c$ with M_c being a very large mass scale of perhaps the order of the Planck mass.*

$Y_u{}^i$ is a D-dimensional Megaverse gauge field equivalent of the universe $Y^\mu(x)$ used in Two Tier renormalization:

$$Y^\mu(z) = z^\mu + i\, Y^\mu(z)/M_c{}^2$$

where $Y^\mu(z)$ is a free QED-like field. The $Y^i(y)$ quantum coordinates will be used in the Megaverse to eliminate potential divergences, in a manner similar to the case of our universe when universe particle interactions are introduced later.

4.2.20.7 Four Types of Fermionic Universe Particles

Assuming universe energies are real-valued,[141] there are four possible types of fermionic universe particles in the Megaverse that are completely analogous to the four species of fermions. Two of these types are tachyonic. It is important to note that DeWitt points out that the Wheeler-DeWitt equation has tachyonic solutions since the mass-like term dependent on $^{(3)}R$ can be positive or negative.[142] A negative mass is an indication of tachyonic behavior wherein the wave propagation of the state functional is not necessarily in time-like directions and is thus tachyonic.

There is a a D-dimensional Dirac equation above. There are three other general types of universe particle equations. (By assumption fermionic universes come in four species like fermions.) The derivation of the four types of universe particles is similar to the derivation of fermion types discussed previously. We will now consider the D-dimensional equivalent for universe particles in the Megaverse.

---

[140] Giving Two-Tier renormalization without infinities.

[141] The energy of universe particles need not be real-valued since universes can 'decay' – unlike elementary particles which are not subject to decay, by definition, since they are assumed to be *fundamental*. We choose to consider the case of universes with real-valued energies. The case of universes with complex-valued energies is a simple extension of the real-value cases considered here.

[142] DeWitt, B. S., Phys. Rev. **160**, 1113 (1967) p. 1124.

The general form of a pure D-dimensional complex Lorentz group[143] boost can be expressed in terms of a complex relative (D – 1)-velocity $\mathbf{v}_c$ between inertial reference frames. A D-dimensional coordinate boost has the form

$$\Lambda_C(\mathbf{v}_c) \equiv \Lambda_C(\omega, \mathbf{v}_c) = \exp[i\omega\hat{\mathbf{w}}\cdot\mathbf{K}]$$

where

$$\omega = (\omega_r^2 - \omega_i^2 + 2i\omega_r\omega_i\, \hat{\mathbf{u}}_r\cdot\hat{\mathbf{u}}_i)^{1/2}$$

and

$$\hat{\mathbf{w}} = (\omega_r\hat{\mathbf{u}}_r + i\omega_i\hat{\mathbf{u}}_i)/\omega$$

with all vectors being (D – 1)-dimensional spatial vectors. We define the real and imaginary unit vectors $\hat{\mathbf{u}}_r\cdot\hat{\mathbf{u}}_r = 1 = \hat{\mathbf{u}}_i\cdot\hat{\mathbf{u}}_i$ with the result

$$\hat{\mathbf{w}}\cdot\hat{\mathbf{w}} = 1$$

The complex relative velocity is

$$\mathbf{v}_c = \hat{\mathbf{w}}\tanh(\omega)$$

The free dynamical equations of the four universe particle species will be generated by D-dimensional Lorentz boosts of the free Dirac equation of a universe particle at rest with the *requirement that the time variable* ($t = y^D$) *and energy are real in the resulting field equations.*[144] The procedure can most easily be performed in D-dimensional momentum space with the Megaverse coordinate space version of the generated equation determined from the momentum space version.

The result is a number of different types of fermion universes that mirror the particle species that we have seen earlier.

- Dirac-like Universe Particle
- Tachyon Universe particle
- Left-Handed Tachyonic Universe Particles
- Right-Handed Tachyonic Universe Particles
- "Up-Quark-like" Universe Particles
- Left-Handed "Down-Quark-like" Tachyonic Universe Particles
- Right-Handed Down-Quark-like Tachyonic Universe Particles

---

[143] The D-dimensional complex Lorentz group has similar features to the 4-dimensional complex Lorentz group. We shall only discuss it to the extent needed for our universe particle type's derivation. See Weinberg (1995) for the 4-dimensional Lorentz group – the D-dimensional Lorentz group generalizes directly from the features of the 4-dimensional Lorentz group.

[144] The D-dimensional "energy" must be real since it relates to the area of the universe – a real number.

Free lagrangians follow for these universe particles which are similar to the corresponding fermion lagragians seen earlier.

4.2.20.8 When Universes Collide: Interactions and Collisions of Universe Particles

4.2.20.8.1 Gravitation and Other Forces

As we saw in Blaha (2015a) the forces involved in the interactions and collisions of universe particles are the force of gravity, and other forces.

We described the baryonic force as an example. The other forces are similar so we will not describe them here. Massless gauge field interactions are of particular interest since they are long range. For example, the normal and Dark baryonic forces are long range forces, both inside universes and in the Megaverse, because their gauge fields are massless.

4.2.20.8.2 Universes in Collision

We assume that the dynamics of universes in collision will be analogous to that of galaxies in collision since gravity is a dominant force in both cases. Colliding galaxies have often been observed. Their dynamics should provide guidance for the case of universes in collision.[145]

It is clear in the case of colliding galaxies, and of colliding large nuclei (gold and lead typically) that there are several types of collisions with differing results. These types of universe collisions can be qualitatively classified as

1. Clean collisions in which universes nudge each other but retain their identity. These are extreme peripheral collisions. If the universes overlap slightly then the typically spherical symmetry of the universes may become distorted and they may become lopsided.[146]

2. Peripheral collisions in which the universes retain their identity but are connected by a trailing string of mass-energy. Eventually the string breaks and the universes separate. Subsequently the pieces of trailing string in each universe contract due to their universe's gravitational effects.

3. Two universes can collide and produce multiple universes.

4. Two universes can collide in a "central" collision and amalgamate into one universe.

---

[145] The high energy collision of atomic nuclei at Brookhaven, CERN and other laboratories also is analogous in overall detail with universes in collision.

[146] The Wilkinson Microwave Anisotropy Probe (WMAP) and the Planck European Space Agency satellite have been accumulating data since 2001 that suggests the universe may be lopsided with hot and cold spots on opposite sides of the universe differing from those on the other side being hotter and colder respectively. *Perhaps the result of a collision when the universe was young.*

We will discuss universe interactions from this viewpoint in more detail later.

4.2.20.9 Bosonic Universe Particles

The previous sections have described fermionic universe particles. In this section we will briefly describe aspects of bosonic (spin 0) universe particles. First it is important to note that the Wheeler-DeWitt equation being second order like the Klein-Gordon equation seems to suggest that universe particles can be bosonic – like Klein-Gordon equation particles.[147] The Wheeler-DeWitt equation has a mass-like term R that can be positive or negative. If the mass term is negative then the wave-like propagation of the state functional (wave equation solution) can be in space-like directions implying a tachyonic solution. Thus the Wheeler-DeWitt equation supports "normal" state functionals that propagate in time-like directions as well as tachyonic propagation.

For this reason we suggest that bosonic universe particles can be either normal or tachyonic. Tachyonic boson universe particles can fission in a manner similar to tachyonic fermionic universe particles. Fermion fission equations also apply to tachyonic bosonic universe particles.

The quantum field theory of normal and tachyonic bosonic universe particles is similar to that of ordinary bosons. See Blaha (2005a) for the boson case discussion which is paralleled by our universe particle formalism.

4.2.20.10 Physical Meaning of Universe Particle Spin

The physical meaning of spin is a continuing discussion topic. We have suggested[148] that spin states are in essence logic states with changes in spin an analogous to changes in logical values in a discourse or computer program. Since the matrix formalism for spin ½ and higher spin states is formally similar to the formalism for angular momentum, one can combine spin and angular momentum as we do in quantum theory.

In the case of universe particles, one can also associate universe particles with "true" and "false" values. Fermionic universe states have $2^{D/2-1}$ truth values and correspond to a multi-valued logic. The numerousness of truth values is due to the D-dimensional space within which universe particles reside.

Naturally one would like to know the physical differences between these $2^{D/2-1}$ types of universe particles. Does the difference reside in different shapes of the universe particles? Or is the difference somehow a consequence of the global mass-energy distribution of the universe that we have not been able to discern since we only know of one universe?

The physical meaning of spin for elementary particles is also somewhat elusive. It does not reflect the flow of charge within a particle. For if it did reflect physical spinning of a

---

[147] One should remember that the Wheeler-DeWitt equation is not in space but in a 6-dimensional manifold, denoted M, of metrics with one "time" dimension – having hyperbolic signature – + + + + + when the metric is positive definite. See DeWitt's paper.

[148] Blaha (2011c) and subsequent books.

particle, the outer edges of a particle such as an electron would be traveling at a speed faster than light. So spin is not a mechanical property of the internal structure of an elementary particle. We have suggested that it is a truth value in the matrix formulation of a 4-valued logic called Asynchronous Logic. Thus it has no certain tangible physical basis.

In the case of universe particles the situation is unclear at present. It could be taken to be an indirect reflection of the structure of mass-energy within a universe. This view would be contrary to our proposed view of elementary particle spin as truth values. So we can only assert that a logic interpretation is the only sensible one (based on our present knowledge or our lack thereof). The physical role of universe particle spin is only evident in interactions between universe particles via gauge fields. Thus one must simply view it as a construct for the present.

Other than our mapping of spin values to logic values in Asynchronous Logic there is little anyone can say about the physical origin of elementary particle spin. Specifying a symmetry group as the origin is not sufficient.

4.2.20.11 Elementary Particles with Time Dependent Masses

The discussions of this section were presented for universe particles. However they could also apply to elementary particles or condensed matter excitations with some changes – primarily changes due to different dimensions.

Elementary particles with time dependent masses do not exist as far as we know. And that is a good thing. The idea of an elementary particle expanding indefinitely, like a universe, would have disastrous consequences for life, as would particles contracting indefinitely. However, particles with oscillating masses would seem to be physically acceptable. The development of an elementary theory with oscillating masses would be an interesting exercise that has applications in condensed matter physics.

4.2.20.12 Impact of Universe Particle Acceleration – Lopsided Internal Structure of Universe

We developed a theory of universe particle interactions. Such interactions would cause universe particles to accelerate and should be detectable within a universe as a "lopsidedness" – there would be a shift of parts of the universe away from the direction of acceleration resulting in a difference in the features of the universe "in front" compared to those "in back" – an acceleration effect just as one sees when a jet accelerates.

Interestingly new data from the Planck observatory of the European Space Agency confirms and extends earlier data from NASA's WMAP observatory that one side of the universe appears different from the other side. There are temperature differences and mass distribution differences – just as one might expect if the universe were accelerating as a unit.

Thus we see the beginning of data suggesting our universe may be moving – "indeed accelerating" – through a Megaverse. Some Planck observatory scientists have suggested their data is a preliminary indication of the Megaverse.

4.2.20.13 Megaverse Baryonic Gauge Field - Plancktons

The conservation of baryon number has been repeatedly investigated by experimenters and found to be true to extremely high accuracy. For decades theorists have suggested that the conservation law follows from the existence of a gauge field in a manner much like electric charge conservation follows from the properties of the electromagnetic abelian gauge field.[149]

We will therefore assume a baryonic gauge field exists that is similar to the electromagnetic field except for features due to its definition and existence in the D-dimensional Megaverse. This field will couple extremely weakly to individual baryons as well as universe particles with non-zero baryon number. We will call this baryonic gauge field particle a *planckton*. Its electromagnetic analogue is the photon. We described D-dimensional baryonic gauge field quantization in Blaha (2015a) and earlier books.

Plancktons propagate in the Megaverse, both within universes, and in the Megaverse external to universes. So we defined the planckton field in D-dimensional Megaverse coordinates. They interact with baryons within a universe with Megaverse coordinates mapped to the curvilinear coordinates in the universe. (This mapping was discussed earlier.)

Since a planckton field in D-dimensional conventional coordinates would lead to divergences we used quantum coordinates:

$$Y^i(y) = y^i + i\, Y_u{}^i(y)/M_u{}^{D/2}$$

with quantum coordinate derivatives defined by

$$\partial_i = \partial/\partial Y^i(y) = \partial/\partial(y^i - Y_u{}^i(y)/M_u{}^{D/2})$$

to obtain a completely finite theory of planckton interactions with elementary particles and universe particles.

We assume plancktons, particle fields, universe fields, gravitation, other gauge fields, and the $Y_u{}^i(y)$ field of quantum coordinates are the only fields in the space between universes in the Megaverse.

However, we believe all gauge fields in the Unified SuperStandard Model have a matching analogue in the Megaverse. These Megaverse fields do not necessarily have any continuity conditions relating them to their counterparts in universes.Universe fields may be confined to universes by surface tension.. Their Megaverse equivalents are spread throughout the Megaverse including inside universes. Megaverse fields must be very small to have avoided detection in astronomical and earth-based experimental studies.

---

[149] See Gell-Mann, M. and Levy, M. *Nuovo Cimento* 16, 705 (1960) for a proof.

4.2.20.14 Beyond the Planckton

4.2.20.14.1 PLANCKTON INTERACTIONS WITH UNIVERSE PARTICLES AND INDIVIDUAL BARYONS

Section 15,1 of Blaha (2015a) describes the second quantization of plancktons in 16 dimensions. In this section we develop an interacting theory in D dimensions of universe particles and plancktons from the model lagrangian terms of universe particles, plancktons and quantum coordinates. We will only consider the case of Dirac-type universe particles since the other cases differ from it only in details. The universe particle – planckton lagrangian terms are:

$$\mathscr{L} = \overline{\psi}(Y(y))[i\gamma^\mu\partial/\partial y^\mu - e_B\gamma^\mu B_{u\mu}(Y(y)) - m(t)]\psi(Y(y)) - \tfrac{1}{4} F_{Bu}{}^{\mu\nu}(Y(y))F_{Bu\mu\nu}(Y(y)) - \\ - \tfrac{1}{4} F_u{}^{\mu\nu}(y)F_{u\mu\nu}(y)$$

where μ, $\nu = 1, 2, \ldots, D$ and where

$$\overline{\psi} = \psi^\dagger\gamma^D$$
$$F_{Bu\mu\nu} = \partial B_{u\mu}(Y(y))/\partial Y^\nu(y) - \partial B_{u\nu}(Y(y))/\partial Y^\mu(y)$$
$$F_{u\mu\nu} = \partial Y_\mu/\partial y^\nu - \partial Y_\nu/\partial y^\mu$$
$$Y^i(y) = y^i + i\, Y_u{}^i(y)/M_u{}^{D/2}$$

$$e_B = e_{B0}/M_u{}^{D/2-2}$$

with e_{B0} a dimensionless coupling constant, and with μ and ν ranging from 1 through D.

The corresponding lagrangian is

$$L = \int d^{(D-1)}y\,\mathscr{L}$$

Note the dimensions of the fields differ in the D dimensional space:

$$Y^\mu \sim [\text{mass}]^{D/2-1}$$
$$B_{u\mu} \sim [\text{mass}]^{D/2-1}$$
$$\psi \sim [\text{mass}]^{(D-1)/2}$$

as can be seen from the above lagrangian as well as earlier equations. Note also that the mass and thus the size of universe particles is time dependent. They can expand or contract with time depending on their internal characteristics (gravitation and effects of elementary particle interactions) which are not embodied in this lagrangian. As a result, this theory, incomplete as it is, does not conserve energy unless m(t) is constant.

The lagrangian generates the baryonic interactions of universe particles using Two-Tier quantum coordinates which prevent infinities in perturbation theory calculations.

The interaction of baryon elementary particles with the baryonic field requires terms in Extended Standard Model covariant derivatives specifying the baryon field interaction baryons with the form

$$e_B\gamma^\mu B_{u\mu}(Y(y))$$

The following sections describe some of the physically significant interactions that the lagrangian implies.

4.2.20.14. 2 CREATION OF UNIVERSES THROUGH GAUGE FIELD FLUCTUATIONS

One of the most exciting questions in Cosmology is the origin of our universe. The conventional view is that it originated in a Big Bang from an infinitesimal point in space. The source of the Big Bang and the prior state of the Cosmos, if there was one, is the subject of much speculation. Based on the particle interpretation of the Wheeler-DeWitt equation, the possibility of a baryonic force strongly supported by conservation of baryon number, and the Megaverse concept, it is reasonable to consider the possibility that the universe originated in a vacuum fluctuation.

In this case there would be two Big Bangs one for our universe and one for an anti-universe. One would expect that they would have opposite corresponding features: one with baryon dominance – one with anti-baryon dominance, and one left-handed – one right-handed.

Our formulation of universe particle theory provides for the generation of a universe particle and anti-particle as a vacuum fluctuation. We view a universe particle as having a substantial excess of baryons, N, as we see in our universe. Its anti-universe at the time of creation (the Big Bang point) is its "mirror image" having the "same" number of anti-baryons (baryon number –N) so that baryon number is conserved by the fluctuation event. Thus the excesses of one universe are compensated by the excesses of the other.

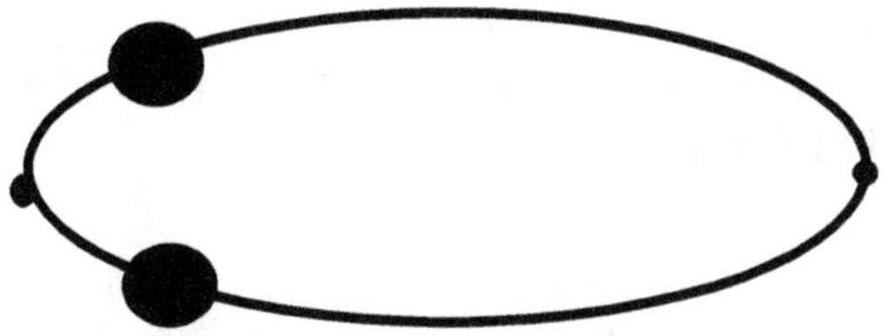

Figure 4.9. Generation of a universe – anti-universe pair as a vacuum fluctuation.

The small value of the coupling constant should lead to an extremely long lifetime for the universes generated by the fluctuation. Thus the 13.7 billion year life of our universe is not unreasonable. The probability of the creation of universes by vacuum fluctuations should be correspondingly small.

4.2.20.14.3 When Universes Collide: Coalescence of Universes

Universes moving in the Megaverse can collide through chance, or due to the planckton field (or gravity or other gauge field interactions) which can cause universes with excess baryons to attract universes with excess anti-baryons.

When universes collide several possibilities present themselves:

1. They can graze each other distorting each other's shape and internal baryon distribution through the baryonic and other forces while maintain their individual identity.

2. They can intermix with both the baryonic, gauge, and gravitational forces causing a redistribution of their masses. They may separate afterwards or may coalesce into a single universe. One result of this may be lopsided universes. Our universe appears to be lopsided. Some cosmologists believe this is due to a near collision of our universe with another shortly after the Big Bang.

In our discussion we have been referring only to the planckton baryon field for the sake of concreteness. But the three other massless particle number fields will also play a role in universe interactions. The relative strength of these interactions is not known since we do not know the size of their coupling constants.

4.2.20.14.4 Fission of Universes

Under certain circumstances the distribution of matter in the universe may lead to the fission of the universe into two separate universes. Our model lagrangian supports this possibility for universe particles. The detailed mechanism of the fission process is not specified by the model.

4.2.20.14.5 Fission of Normal universes

The fission of universe particles in our universe particle model is depicted in the Feynman diagram in Fig. 4.10,

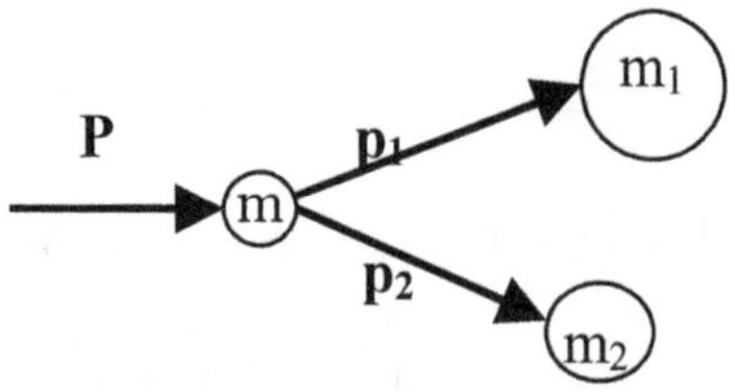

Figure 4.10. Fission of a universe particle into two universe particles.

The sum of the masses of the output universe particles is usually less than the original universe particle mass. However if the fission takes a long time and the masses are time dependent then the produced universe particles combined masses may exceed the original universe's mass.

4.2.20.14.6 TACHYON UNIVERSE PARTICLE FISSION TO MORE MASSIVE UNIVERSE PARTICLES

In Blaha (2007a) we showed that a tachyonic (faster than light) particle could fission into particles of larger mass through the conversion of momentum into mass. In this section we show that a tachyonic universe particle may fission into two more massive universe particles.[150] This phenomenon is of particular interest because it enables tachyonic universes to spawn in a new novel way not previously considered in discussions of the origin of universes.

A simple model lagrangian for a tachyonic universe particle is

$$\mathcal{L}_{\|} = \overline{\psi}_T{}^S(Y(y))[\gamma^\mu\partial/\partial y^\mu - e_B\gamma^\mu B_{u\mu}(Y(y)) - m(t)]\psi(Y(y)) - \tfrac{1}{4}F_{Bu}{}^{\mu\nu}(Y(y))F_{Bu\mu\nu}(Y(y)) - $$
$$- \tfrac{1}{4}F_u{}^{\mu\nu}(y)F_{u\mu\nu}(y)$$

We assume m(t) is constant.

When a particle or a universe particle fissions (decays) one normally expects that the masses of the particles or universe particles produced by the decay to be smaller than the mass of the original particle or nucleus. In the case of tachyonic (faster-than-light) elementary particles or universe particles a much different possibility is present: a tachyon can decay into heavier tachyons. We will consider the specific case of a tachyon universe particle decaying into two universe particles whose total mass is greater than the original. (See Fig. 4.11.)

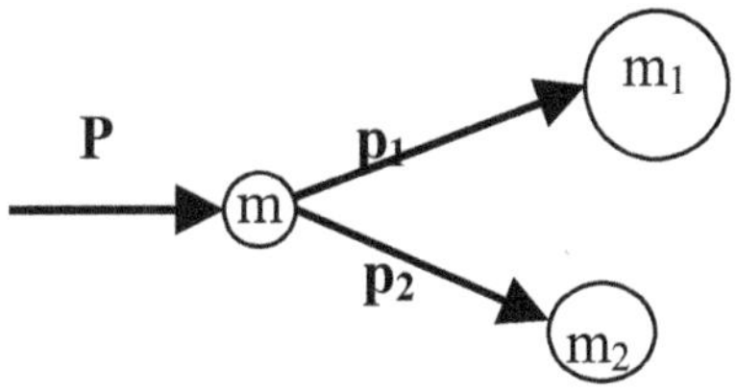

Figure 4.11. Two universe particle decay of a tachyon universe particle.

We will assume the initial tachyon universe particle has zero energy ($p^D = 0$) and thus the tachyons universe particles emerging from the decay also have total universe particle energy zero. The analysis is based on conservation of total universe energy and momentum in Megaverse space outside of universes. The below discussion applies to D-dimensional space with (D – 1)-dimensional spatial coordinates.

Momentum conservation implies

[150] We will use the term mass here to denote mass-energy. Since we identified mass as a multiple of area earlier the comments here would appear to apply to universe area as well.

$$\mathbf{P} = \mathbf{p_1} + \mathbf{p_2}$$

Since all energies are zero

$$(cP)^2 = (c\mathbf{P})^2 = m^2$$
$$(cp_1)^2 = (c\mathbf{p_1})^2 = m_1^2$$
$$(cp_2)^2 = (c\mathbf{p_2})^2 = m_2^2$$

where $P = |\mathbf{P}|$, $p_1 = |\mathbf{p_1}|$, and $p_2 = |\mathbf{p_2}|$. If we now square the above equation for **P** and then use the above three equations we obtain

$$m^2 = m_1^2 + m_2^2 + 2m_1m_2 \cos\theta$$

where θ is the opening angle between the emerging universe particles momenta $\mathbf{p_1}$ and $\mathbf{p_2}$. There are a number of interesting cases:

Case $\theta = 0$:

$$m = m_1 + m_2$$

The masses of the outgoing universe particles sum to the mass of the original tachyon universe particle.

Case $\theta = \pi/2$:

$$m^2 = m_1^2 + m_2^2$$

The masses of each outgoing universe particle tachyon are less than the mass of the original tachyon universe particle.

Case $\theta = \pi$:

$$m^2 = (m_1 - m_2)^2$$

In this case either $m_1 > m$ or $m_2 > m$. Thus one of the outgoing tachyon universe particles has a greater mass than the original tachyon universe particle. Mass is effectively created from the spatial momentum of the initial universe particle. This process is the inverse of normal particle and universe particle fission where the sum of the outgoing masses is always less than the original particle's mass and the difference is mass converted into energy in the form of additional photons.

This last case, where one of the outgoing universe particles is more massive than the original universe particle, is not just for $\theta = \pi$. Since

$$\cos\theta = (m^2 - m_1^2 - m_2^2)/(2m_1m_2)$$

we see that the sum of the outgoing universe particle masses is always greater than the original tachyon universe particle *mass (except when* $\theta = 0$*)* since

$$\cos\theta = 1 + [m^2 - (m_1 + m_2)^2]/(2m_1m_2) \le 1$$

and thus

$$[m^2 - (m_1 + m_2)^2]/(2m_1m_2) \le 0$$

Note $m = m_1 + m_2$ only if $\theta = 0$.

Since we can transform the above discussion to the case of universe particle tachyons having non-zero Megaverse energy using an ordinary D-dimensional Lorentz transformation the discussion in this subsection is general.

We therefore conclude that when a tachyon universe particle decays into two tachyon universe particles the sum of the masses of the produced tachyon universe particles is greater than the mass of the original tachyon universe particle except if the angle between the momenta of the produced tachyon universe particles is zero. In that case the sum of the masses of the produced tachyon equals the mass of the original tachyon universe particle and the produced universe particles overlap.

4.2.20.14.7 UNIVERSE PARTICLE – PLANCKTON INTERACTIONS

These interactions are quite similar to Two-Tier electromagnetic interactions except that universe particles have time-dependent masses, and that the space is D-dimensional.

The interactions have a new aspect due to the time dependence of the universe particle masses. This feature is illustrated by Fig. 4.12: the mass of a universe particle after a baryonic interaction vertex is the same as it was before the interaction assuming the point-like interaction specified in the lagrangian.

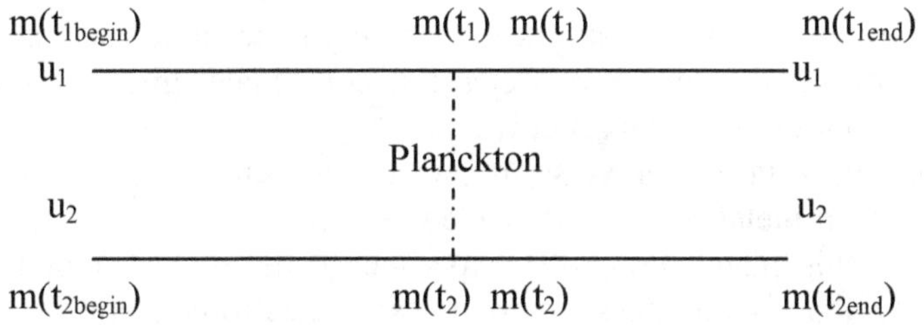

Figure 4.12. A Feynman diagram illustrating the continuity of a universe particle mass through a Planckton interaction.

The reader may verify this by writing the perturbation theory equivalent. A universe particle vertex corresponds to

$$iS_F^{TT}(y_1, y_2)\gamma^\mu iS_F^{TT}(y_2, y_3)$$

The universe particle mass is the same on either side of the interaction vertex.

4.2.20.15 Internal Structure of Universe Particles

We have developed a beginning in planckton field theory that gives interactions between baryons. This theory is applicable to universe-universe interactions. It also yields baryon particle – baryon particle interactions as well as baryon particle – universe particle interactions.

It is possible for a planckton to be emitted in one universe and interact with a baryon elementary particle in another universe. This type of "probe" must be a high energy probe just as a photon probe of the internal structure of a nucleon[151] must be a high energy photon to bring out the nucleon's internal structure (parton model).

In this section we will discuss planckton probes of other universes, and the internal structure of a universe as a mass distribution governed by gravitation as it relates to universe particles.

4.2.20.16.1 PLANCKTON PROBES

Plancktons can be generated in one universe and be used to probe the baryon distribution of another universe. Since the planckton propagator is expressed in Megaverse coordinates the baryon distribution in the target universe will be a distribution in Megaverse coordinates. Megaverse coordinates can be expressed in terms of the curved space-time coordinates of a universe x^μ. However the inversion of the map between universe coordinates and Megaverse coordinates

$$x^\mu = f^{-1\mu}(y)$$

is not 1:1 since x^μ is 4-dimensional and y is a D-dimensional vector. The universe coordinates x^μ are each individually determined up to a subspace. One might be concerned about this situation but the determination of the distribution in Megaverse coordinates gives a more direct picture not convoluted by the curvature of the target universe.

The detailed probing of a target universe requires high energy plancktons. The similarity of this procedure to deep inelastic electron-nucleon scattering is obvious to the high energy physicist. But in doing this planckton probe experiment one obtains a picture of a different universe – something that is not possible to do with electromagnetic or graviton probes.

4.2.20.16.2 INTERNAL STRUCTURE OF A UNIVERSE PARTICLE

The development of the theory of universe particles which resulted in the lagrangian appearing earlier does not fully describe universe particles since it neglects the internal structure

[151] Deep inelastic electron-nucleon scattering.

of a universe particle. The internal structure of a universe particle is primarily determined by gravitation, electromagnetic effects and nuclear physics.

Consequently the full lagrangian of a universe particle has the form

$$\mathcal{L}_{tot} = \mathcal{L}_{internal} + \mathcal{L}$$

where $\mathcal{L}$ is determined above. As a result the complete quantum wave function of a universe particle has the form

$$\psi_{tot} = \psi_{internal}(Y)\psi_{ext}(Y)$$

where $\psi_{internal}(Y)$ is the internal wave function and $\psi_{ext}(Y)$ is external wave function. It seems reasonable to have a separable equation except when universes collide. In that situation a perturbative mixing of the universes and their wave functions applies and it may be possible to calculate the collision output universes by introducing a further interaction between the internal and external aspects of the universe particles.

4.2.21 Megaverse SuperSymmetry

The discussion of the Θ-symmetry and SuperSymmetry in our 4-dimensional space-time was presented earlier in section 4.2.17. A similar discussion of Θ-symmetry and SuperSymmetry with additional ramifications will now be considered. It was discussed to some extent in Blaha (2017c).

An important new feature relating to Θ-symmetry and SuperSymmetry is the dimension of the Megaverse D = 192. Due to the value of this dimension we can define a U(192) unitary Reality group that transforms the complex-valued coordinates of the Megaverse to real-valued coordinates. We also have the Θ-symmetry for interactions. In addition, we can define a 192-dimensional Lorentz group that transforms between physical reference frames. Lastly we can define 192 antisymmetry coordinates that supports Supersymmetric field theory in the Megaverse.

This section discusses these possibilities. Much of the discussion parallels the discussions in Blaha (2017c).

4.2.21.1 Θ-SymmetryTransformations in the Megaverse

The choice of dimension, 192, enables us to extend the Θ-Symmetry toMegaverse coordinates. Let $C_\Theta(y)$ be a local U(4) Θ-Symmetry transformation that rotates interactions in Megarverse space y. Then applying a Θ-Symmetry transformation to the Dirac equation in Megaverse space:[152]

$$\gamma_\mu D^\mu\psi \ = \ \gamma_\mu\{\partial^\mu + i\ [g_\Theta A_\Theta{}^{1\mu}(y) + g_\Theta A_\Theta{}^{2\mu}(y) + \mathbf{A}_I{}^{1\mu}(y) + \mathbf{A}_I{}^{2\mu}(y)]\}\psi(y) = 0$$

we find it transforms to

$$\gamma_\mu\{\partial^\mu + i\ [g_\Theta A'_\Theta{}^{1\mu}(y) + g_\Theta A'_\Theta{}^{2\mu}(y) + \mathbf{A'}_I{}^{1\mu}(y) + \mathbf{A'}_I{}^{2\mu}(y)]\}\psi'(y') = 0$$

where

$$\begin{aligned}
A'_\Theta{}^{1\mu}(y') &= C_\Theta(y)A_\Theta{}^{1\mu}(y)C_\Theta{}^{-1}(y) - i\ C_\Theta(y)\partial^\mu C_\Theta{}^{-1}(y)/g_\Theta \\
A'_\Theta{}^{2\mu}(y') &= C_\Theta(y)A_\Theta{}^{2\mu}(y)C_\Theta{}^{-1}(y) \\
\mathbf{A'}_I{}^{1\mu}(y') &= C_\Theta(y)\mathbf{A}_I{}^{1\mu}(y)\ C_\Theta{}^{-1}(y) \\
\mathbf{A'}_I{}^{2\mu}(y') &= C_\Theta(y)\mathbf{A}_I{}^{2\mu}(y)\ C_\Theta{}^{-1}(y) \\
\psi'(y') &= C_\Theta(y)\psi(y)
\end{aligned}$$

The Θ-Symmetry interaction is experienced by all particles (fields) in the Megaverse. It is also experienced by all universes in the Megaverse and thus affects universe dynamics.

[152] We omit the Species group interaction.

4.2.21.2 Megaverse Lorentz Group

The Megaverse is a Euclidean space with 192 complex-valued dimensions. We can choose[153] one dimension to be the 'time' coordinate since we want to have a dynamics for physical phenomena. We chose the real-valued $-iy^{192}$ coordinate to be the time coordinate. Then the Megaverse has a Complex Lorentz group with one time coordinate and 191 spatial coordinates.

Dynamics is invariant under Megaverse Complex Lorentz group transformations. The discussion above and Blaha (2017c) describe the possible motions of universes, and universe interactions, in detail.

4.2.21.3 Megaverse SuperSpace

The Megaverse offers an exciting new platform for SuperSymmetry studies of the Unified SuperStandard Model and possible SuperSymmetric generalizations using Megaverse Superspace with antisymmetric coordinates in addition to normal coordinates.

We can define a generalized set of Megaverse coordinates that associate 192 antisymmetric coordinates with the 192 normal Megaverse coordinates:

$$\{y^\mu, \Theta^\nu\}$$

where $\mu = 1, \dots, 192$ *and* also $\nu = 1, \dots, 192$. We now have an equal numbers of normal and antisymmetric coordinates in Megaverse superspace. This was not the case in 4-dimensional space-time.

Supersymmetric transformations on superspace coordinates have the infinitesimal form:

$$y^\mu \to y^\mu + i\,\varepsilon\,\Gamma^\mu\Theta$$
$$\Theta^\nu \to \Theta^\nu + \varepsilon^\nu$$

where Γ^μ is a 192×192 matrix generalization of the Dirac γ matrices.[154] There are 192 Γ matrices.

Using Megaverse superspace coordinates one can proceed to construct a superfields formalism with many features analogous to the 4-dimensional formulation but of significantly greater complexity. We defer doing this in view of the finiteness of Megaverse Two-Tier Quantum Field Theory which, like the 4-dimensional case, yields totally finite results in all orders of perturbation theory. We note the major benefit of 4-dimensional SuperField Theory was the finiteness of calculations in perturbation theory. Given the complexity of Megaverse

---

[153] This can be done by a U(192) transformation applied to the coordinates of a 'physical' reference frame yielding one coordinate with a purely imaginary values. Then the Euclidean metric becomes the Lorentz metric in the transformed frame. All other physical reference frames are defined to be Complex Lorentz transformations of this coordinate system.

[154] See Blaha (2017c) for their definition for arbitrary dimension D.

SuperField perturbation theory it would appear the simplicity of Two-Tier Quantum Field Theory is preferable.

4.3 An Axiomatic Unified SuperStandard Model

In this chapter we have provided a derivation of the Unified SuperStandard Model in 4-dimensional space-time and the Megaverse. The reader will note that it is much in the manner of Euclid's derivation of Geometry with axioms and 'theorems' together with supplementary assumptions developed 'along the way.' The basis of the derivation in Logic and the Complex Lorentz group and General Relativity strongly argues for its correctness given the state of our knowledge of elementary particles, their interactions, and gravity. Is elementary particle Theory then finished? We believe not. First there are the open questions of the origin of mass values and coupling constants in the theory.. Next there is the question of fine points of the theory such as CP violation as well as significant cosmological open questions.

Thirdly, there is the question of a deeper basis for the theory. At the moment the only approach that we have available theoretically is the consideration of all the other possible theories that one might construct of elementary particles and gravitation. We circumvented this issue by assuming the basis of the theory is in Logic, the Complex Lorentz group and General Relativity. It is not clear that any other approach is feasible. On these grounds we rest our case.

The theory presented is clear, mathematically correct, and, in our view, meets Parmenides specification of unchanging Reality. In th next chapter we address how and why it maps to the physical world.

5. Unified SuperStandard Model Map to Reality

In chapter 4 we derived a finite Unified SuperStandard Model from basic assumptions. This derivation gave us an unchanging theory of Reality conceptually similar to the concept of Parmenides. The question now becomes: How is that Reality imposed on the material universe that we see – everywhere with 'infinite precision?' There appear to be three possible answers to that question:

1. Our primitives, axioms and derivation are the only possible choices (modulo minor variations). They are naturally adapted to describe physical reality by construction. This view is similar to that of Professor Hawking who said he did not see a necessity for God in the workings of the universe.

2. There are other possible explanations of the features of the material universe such as a SuperString theory. These explanation(s) may lead to our theory.

3. There is an 'Unmoved Mover' that causes Reality (our theory) to be imposed on the material universe (and Megaverse.) Some see this entity as God.

In any case it is clear that our theory maps directly to the material universe that we see although there are questions that remain to be addressed.

Appendix A. The Local Definition of Asymptotic Particle States

This appendix reprints S. Blaha, "The Local Definition of Asymptotic Particle States", IL Nuovo Cimento **49A**, 35 (1979).[155] It describes the PseudoQuantization of boson and fermion field theories for use in the quantization of fields in universes and the Megaverse in chapter 7.

[155] © Copyright Stephen Blaha 1978.

The Local Definition of Asymptotic Particle States (*).

S. BLAHA

Physics Department, Williams College - Williamstown, Ma. 01267

(ricevuto il 28 Luglio 1978)

Summary. — A generalization of quantum field theory is described which has a unique particle interpretation even in space-times where no global timelike co-ordinate exists. The formulation is described in detail for the case of scalar bosons and spin-one-half fermions in flat space-time. We show that it is possible to construct a model in our approach which is physically equivalent to any given model in the usual formulation. In addition, a new class of models can be constructed which are not possible in the usual formulation. This class includes quantum action-at-a-distance models which can be used to develop models with higher-derivative field equations which are unitary. Our formulation allows some latitude in the choice of boundary conditions, so that one can opt for a continuum of possible Green's functions ranging from Feynman propagators to principal-value propagators (half advanced-half retarded).

1. – Introduction.

Our experience in flat space-time has fostered the opinion that a given action leads to a unique quantum field theory upon implementation of the canonical quantization procedure. This is apparently not true in general. A given action corresponds to an infinity of physically inequivalent quantum field theories in nonstatic space-times where no timelike Killing vector exists [1,2]. The origin of this plurality of quantum theories can be seen in free field the-

(*) Supported in part by grants from the National Science Foundation, and Research Corporation.

(1) S. A. FULLING: *Phys. Rev. D*, **7**, 2850 (1973); C. SOMMERFIELD: *Ann. Phys.*, **84**, 285 (1974).

(2) B. DEWITT: *Phys. Rep.*, **19**, 295 (1975).

ories (cf. FULLING [1]). The usual quantization procedure is based on a definition of positive frequency which selects an acceptable complete orthonormal set of field equation solutions to use in field quantization. In nonstatic space-times no unique criterion exists for defining positive frequency. As a result, there is no restriction on the choice of complete orthonormal set of field equation solutions used to Fourier-expand fields. Having different choices leads to unitarily (and physically) inequivalent representations of the field algebra. The set of physical particle states in one quantization is generally not unitarily related to the set of physical states in another quantization [1].

The absence of a criterion to select the « correct » quantum-field theory in the usual formulation has led us to consider a generalization of quantum field theory. In this generalization we introduce extra degrees of freedom in such a way that quantizations based on differing definitions of positive frequency are unitarily equivalent. Thus for a given action there is one resulting quantum field theory up to unitary equivalence.

In particular the physical particle states of different quantizations are related by a unitary transformation. Since the particle number operator is invariant under this transformation, a N-particle state in one quantization is a superposition of N-particle states in any other quantization. This is made possible by a local definition of particle states in the Fourier-transformed space (momentum space in the case of flat space-time).

It is important to note that the plethora of inequivalent quantizations in the usual formulation is faced by *one* observer. It is not a question of quantizations in different co-ordinate systems corresponding to different observers. The differences in the quantizations of two relatively accelerating observers, for example, are physically real and, in fact, also exist within the framework of our formulation. Relatively accelerating observers will, in general, « see » different numbers of particles.

Sections **2** and **3** contain our formulation of a free-scalar-boson field theory and a free spin-one-half fermion field theory in flat space-time. Significant differences exist between our formulation and the usual formulation. However, models exist in our formulation which make predictions which are identical to those of conventional field theory models, *e.g.*, quantum electrodynamics. Models also exist in our formulation which are completely outside the framework of the usual formulation. For example, a choice of boundary conditions is possible in our formulation which allows for virtual particles to propagate via non-Feynman propagators. In general, our particle propagator has the form

$$G = \sin^2\theta\, G_F + \cos^2\theta\, C G_F^* C^{-1}\,, \tag{1}$$

where θ is an arbitrary angle, G_F the usual Feynman propagator with G_F^* its complex conjugate, and where C is the relevant charge conjugation matrix. G is a Feynman propagator if $\theta = \pi/2$.

If $\theta = \pi/4$ then G is a principal-value propagator (half advanced-half retarded). This type of Green's function has appeared in classical action-at-a-distance theories. Our formulation thus encompases quantum action-at-a-distance theories. The use of principal-value propagators allows a substantial enlargement of the class of unitary, renormalizable field theory models. For example, models with higher-derivative field equations cannot simultaneously satisfy the requirements of positive probabilities and unitarity, if Feynman propagators are used. But if principal-value propagators are used, both requirements can be consistently satisfied (3). This has allowed us to previously construct a unitary, higher-derivative non-Abelian model of the strong interactions with a manifest linear potential and quark confinement (4). Of course, the use of principal-value propagators leads to a different type of analytic structure for amplitudes. We take the view that analyticity is an experimental question rather than a fundamental requirement on field theory (5). It is amusing to note that confinement of color in this model serves to sharply dampen if not eliminate the potential nonanalyticity.

We will discuss our formulation of non-Abelian field theories in detail in a subsequent paper (6).

2. – Boson quantization.

In flat space-time a timelike co-ordinate exists and as a result the Hamiltonian occupies a privileged position in defining positive frequency. In a non-static space-time, with no global timelike co-ordinate, no corresponding operator exists and the definition of « positive frequency » appears to be arbitrary. Consider a free scalar-field theory in such a situation. The field equation has an infinite number of possible complete orthonormal sets of solutions which span the space of solutions. Consider two possible sets: $\{\chi_\alpha, \chi_\alpha^*\}$ and $\{\psi_\beta, \psi_\beta^*\}$, where the χ_α are positive frequency with respect to one definition of positive frequency, and ψ_β are positive frequency with respect to a different definition. Then mode expansions of the scalar field

$$\varphi(x) = \sum_\alpha [\chi_\alpha(x) A_\alpha + \chi_\alpha^*(x) A_\alpha^\dagger], \tag{2}$$

$$\varphi(x) = \sum_\beta [\psi_\beta(x) b_\beta + \psi_\beta^*(x) b_\beta^\dagger], \tag{3}$$

(3) S. Blaha: *Phys. Rev. D*, **10**, 4268 (1974).
(4) S. Blaha: *Phys. Rev. D*, **11**, 2921 (1975).
(5) R. E. Cutkowsky, P. V. Landshoff, D. I. Olive and J. C. Polkinghorne: *Nucl. Phys.*, **12** B, 281 (1969); T. D. Lee: in *Quanta-Essays in Theoretical Physics Dedicated to Gregor Wentzel*, edited by P. G. O. Freund, C. J. Gobel and Y. Nambu (Chicago, Ill., 1970); H. Rechenberg and E. C. G. Sudarsham: *Nuovo Cimento*, **14** A, 299 (1973).
(6) S. Blaha: *Nuovo Cimento*, **49** A, 58 (1978).

can be inverted to relate the Fourier coefficient operators

$$A_\alpha = \sum_\beta [C_{\alpha\beta} b_\beta + \tilde{C}_{\alpha\beta} b_\beta^\dagger] , \tag{4}$$

where $C_{\alpha\beta}$ and $\tilde{C}_{\alpha\beta}$ are c-number functions of α and β. Equation (4) shows that A_α is related to b_β and $b_\beta^\dagger$ through a local Bogoliubov transformation. As a result, the quantizations are, in general, not unitarily equivalent, have different vacua, and different particle interpretations [1,2]. The basis of this difficulty is the noncommutativity of Fourier coefficient operators and their Hermitian conjugates

$$[b_\beta, b_{\beta'}^\dagger] = \delta_{\beta\beta'} . \tag{5}$$

We shall propose a generalization of quantum field theory in which (in the free-field case) the Fourier coefficient operators and their Hermitian conjugates commute. In order to maintain the quantum character of the theory a supplementary field and the corresponding Fourier coefficient operators will be introduced. We shall confine our discussion to flat space-time in this section, and in sect. **3** which deals with free spin-one-half fermion quantization. In sect. **4** we discuss the particle interpretation of the formulation in nonstatic space-time.

Let us provisionally introduce the Lagrangian

$$\mathscr{L} = \partial_\mu \varphi_1 \partial^\mu \varphi_2 - \tfrac{1}{2} \partial_\mu \varphi_1 \partial^\mu \varphi_1 - m^2 \varphi_1 \varphi_2 + \tfrac{1}{2} m^2 \varphi_1^2 . \tag{6}$$

Following canonical procedures, we obtain the field equations

$$(\Box + m^2)\varphi_i = 0 , \tag{7}$$

for $i = 1, 2$ and the canonical momenta

$$\pi_1 = \dot{\varphi}_2 - \dot{\varphi}_1 \tag{8}$$

and

$$\pi_2 = \dot{\varphi}_1 , \tag{9}$$

which are taken to satisfy the canonical equal-time commutation relations

$$[\varphi_i(x), \pi_j(y)] = i\delta_{ij} \delta^3(\boldsymbol{x} - \boldsymbol{y}) , \tag{10}$$

$$[\varphi_i(x), \varphi_j(y)] = [\pi_i(x), \pi_j(y)] = 0 , \tag{11}$$

for $i, j = 1, 2$. Equations (9) and (10) imply

$$[\varphi_1(x), \dot{\varphi}_1(y)] = 0\,, \tag{12}$$

$$[\varphi_1(x), \dot{\varphi}_2(y)] = i\delta^3(\boldsymbol{x} - \boldsymbol{y})\,, \tag{13}$$

$$[\varphi_2(x), \dot{\varphi}_2(y)] = i\delta^3(\boldsymbol{x} - \boldsymbol{y})\,, \tag{14}$$

at equal times. The most general form for the mode expansion of the fields is

$$\varphi_1(x) = \int d^3k\,[(C_{11}A_{1k} + C_{12}A_{2k})f_k(x) + (\bar{C}_{11}A^\dagger_{1k} + \bar{C}_{12}A^\dagger_{2k})f^*_k(x)]\,, \tag{15}$$

$$\varphi_2(x) = \int d^3k\,[(C_{21}A_{1k} + C_{22}A_{2k})f_k(x) + (\bar{C}_{21}A^\dagger_{1k} + \bar{C}_{22}A^\dagger_{2k})f^*_k(x)]\,, \tag{16}$$

where $(2\pi)^{\frac{3}{2}}(2\omega_k)^{\frac{1}{2}}f_k(x) = \exp[-ik\cdot x]$ and where C_{ij} and $\bar{C}_{ij}$ are a set of constants. In view of the afore-mentioned difficulties stemming from the non-commutativity of a Fourier-coefficient operator and its Hermitian conjugate, we are led to impose the commutation relations

$$[A_{ik}, A_{jk'}] = [A^\dagger_{ik}, A^\dagger_{jk'}] = 0 \tag{17}$$

and

$$[A_{ik}, A^\dagger_{jk'}] = (1 - \delta_{ij})\,\delta^3(\boldsymbol{k} - \boldsymbol{k}')\,, \tag{18}$$

for $i, j = 1, 2$. We define two vacua (which are in fact related) $|0\rangle_1$ and $|0\rangle_2$ by

$$A_{1k}|0\rangle_2 = A^\dagger_{1k}|0\rangle_2 = 0\,, \tag{19}$$

$$A_{2k}|0\rangle_1 = A^\dagger_{2k}|0\rangle_1 = 0 \tag{20}$$

with

$$A_{2k}|0\rangle_2 \neq 0\,, \quad A^\dagger_{2k}|0\rangle_2 \neq 0\,, \tag{21}$$

and

$$A_{1k}|0\rangle_1 \neq 0\,, \quad A^\dagger_{1k}|0\rangle_1 \neq 0\,, \tag{22}$$

for all k. These definitions are motivated by the need for vacua which would be invariant under Bogoliubov transformations—a necessary requirement if the difficulties of particle interpretation caused by relations of the form of eq. (4) are to be avoided. Let us define the local Bogoliubov transformation

$$A_{ik}(\lambda_1, \lambda_2) = B_{\lambda_1\lambda_2}A_{ik}B^{-1}_{\lambda_1\lambda_2} = \\ = \exp[i\lambda_1]\cosh\lambda_2 A_{ik} + \exp[-i\lambda_1]\sinh\lambda_2 A^\dagger_{ik}\,, \tag{23}$$

where λ_1 and λ_2 are functions of the momentum k. The operator B has the form

$$B_{\lambda_1\lambda_2} = \exp\left[2i\int d^3k\lambda_1(k)\Gamma_{3k}\right]\exp\left[2i\int d^3k\lambda_2(k)\Gamma_{2k}\right], \tag{24}$$

where

$$\Gamma_{3k} = (A^\dagger_{2k}A_{1k} + A_{2k}A^\dagger_{1k})/2\,, \tag{25}$$

$$\Gamma_{2k} = i(A^\dagger_{2k}A^\dagger_{1k} - A_{2k}A_{1k})/2\,. \tag{26}$$

If we also define

$$\Gamma_{1k} = -(A^\dagger_{2k}A^\dagger_{1k} + A_{2k}A_{1k})/2\,, \tag{27}$$

then these operators satisfy the commutation relations of a $SU_{1,1}$ algebra:

$$[\Gamma_{1k}, \Gamma_{2k'}] = -i\delta_{kk'}\Gamma_{3k}\,, \quad [\Gamma_{2k}, \Gamma_{3k'}] = i\delta_{kk'}\Gamma_{1k}\,, \quad [\Gamma_{3k}, \Gamma_{1k'}] = i\delta_{kk'}\Gamma_{2k}\,. \tag{28}$$

Thus the group of local Bogoliubov transformations is an infinite tensor product of $SU_{1,1}$ groups. It should be noted that $|0\rangle_2$ and $|0\rangle_1$ are invariant under this group. The equations of motion and equal-time commutation relations are also invariant under this group. These properties will enable us to show the uniqueness of the particle interpretation of our formulation in nonstatic space-time in sect. **4**. The Casimir operator for the k-th $SU_{1,1}$ algebra,

$$\Gamma^2_k = \Gamma^2_{3k} - \Gamma^2_{1k} - \Gamma^2_{2k}\,, \tag{29}$$

$$\Gamma^2_k = N_k(N_k + 2)\,, \tag{30}$$

allows us to identify the particle number operator (cf. sect. **4** below for its derivation)

$$N = \int d^3kN_k = \int d^3k[A^\dagger_{2k}A_{1k} - A_{2k}A^\dagger_{1k}]\,, \tag{31}$$

which is left invariant by the Bogoliubov transformations.

If we compare our formulation to the usual one at this stage, we see that the enlargement of the field algebra has allowed us to define a group of local Bogoliubov transformations which is unitary and leaves the vacuum invariant —two properties not possible in the usual approach.

We now define inner products in our formalism. The structure of the commutation relations eq. (17) and (18) together with the nature of the defined vacua suggest that kets can be taken to have the form

$$|\alpha\rangle = A^\dagger_{2k_1}A^\dagger_{2k_2}\ldots A_{2q_1}A_{2q_2}\ldots|0\rangle_2 \tag{32}$$

and that bras should have the form

$$\langle\alpha| = {}_1\langle 0|A_{1k_1}A_{1k_2}\ldots A^\dagger_{1q_1}A^\dagger_{1q_2}\ldots\,. \tag{33}$$

(We could have chosen to construct kets using $|0\rangle_1$ and bras using ${}_2\langle 0|$ with no change in consequences.) The form of the commutation relations (which are used to reduce the inner products to a multiple of ${}_1\langle 0|0\rangle_2$) imply that the dual of the ket space is not its Hermitian conjugate. In our case the algebra reduces inner products to ${}_1\langle 0|0\rangle_2$ which we define to be unity.

We can relate the dual of a ket to its Hermitian conjugate through the introduction of a Dirac metric operator [8]. We define the operator, γ, by

$$\gamma^{-1}A_{1k}\gamma = A_{2k}, \qquad \gamma^{-1}A_{2k}\gamma = A_{1k}, \tag{34}$$

$$\gamma|0\rangle_1 = |0\rangle_2 . \tag{35}$$

We find

$$\gamma = \exp\left[-\frac{i\pi}{2}\int \mathrm{d}^3k\,[A_{2k}^\dagger A_{2k} + A_{1k}^\dagger A_{1k} - A_{2k}^\dagger A_{1k} - A_{2k}A_{1k}^\dagger]\right], \tag{36}$$

which implies that γ satisfies the necessary conditions for a metric operator, $\gamma = \gamma^\dagger = \gamma^{-1}$. The norm of a state $|\alpha\rangle$ can thus be defined by

$$(|\alpha\rangle)^\dagger\gamma|\alpha\rangle \tag{37a}$$

and inner products will generally have the form

$$(|\beta\rangle)^\dagger\gamma|\alpha\rangle . \tag{37b}$$

The adjoint operator is defined by

$$A^* = \gamma^{-1}A^\dagger\gamma . \tag{38}$$

Physical observables must be self-adjoint, $A^* = A$. Self-adjoint operators play the same role as Hermitian operators do in the usual formulation. In particular the Hamiltonian must be self-adjoint, if we are to have conservation of norm. Because φ_2 satisfies a Jordan-Pauli commutation relation, we shall introduce interactions in our model using only $\varphi_2(x)$. As a result φ_2 must also be self-adjoint if the Hamiltonian is to be self-adjoint.

---

(7) Earlier two field formalisms have been considered by G. MIE: *Ann. Phys. Lpz.*, **37**, 511 (1912); P. A. M. DIRAC: *Comm. Dublin Inst. Advanced Studies*, **180** A, 1 (1942); W. PAULI: *Rev. Mod. Phys.*, **15**, 175 (1943); M. FROISSART: *Suppl. Nuovo Cimento*, **14**, 197 (1959); T. D. LEE and G. C. WICK: *Phys. Rev. D*, **2**, 1033 (1970). Our motivation and formulation differ substantially from them. Ref. (3,4) above do describe models which can be directly incorporated within the framework of our formulation.

(8) W. PAULI: *Rev. Mod. Phys.*, **15**, 175 (1943).

The energy-momentum tensor is defined by

$$T^{\mu\nu} = -g^{\mu\nu}\mathscr{L} + \frac{\delta\mathscr{L}}{\delta\partial_\mu\varphi_1}\partial^\nu\varphi_1 + \frac{\delta\mathscr{L}}{\delta\partial_\mu\varphi_2}\partial^\mu\varphi_2 \tag{39}$$

with the Hamiltonian given by the 0-0 component. It is easy to verify that the requirements of Poincaré invariance, and the Schwinger commutation relations for $T^{\mu\nu}$ are met.

The equal-time commutation relations, and the self-adjointness of H and φ_2 place six constraints on the constants C_{ij} and $\tilde{C}_{ij}$ in eqs. (15) and (16). After some algebra we find that we are able to express the field operators in the form

$$\varphi_1(x) = \int d^3k\left[\left(\frac{\cos(\theta_1-\theta_2)}{\sin\theta_1}A_{1k} + \frac{\sin(\theta_1-\theta_2)}{\sin\theta_1}A_{2k}\right)f_k(x) + \left(\frac{\cos(\theta_1-\theta_2)}{\cos\theta_1}A^\dagger_{1k} - \frac{\sin(\theta_1-\theta_2)}{\cos\theta_1}A^\dagger_{2k}\right)f^*_k(x)\right], \tag{40}$$

$$\varphi_2(x) = \int d^3k\,[(\cos\theta_2 A_{2k} + \sin\theta_2 A_{1k})f_k(x) + (\sin\theta_2 A^\dagger_{2k} - \cos\theta_2 A^\dagger_{1k})f^*_k(x)], \tag{41}$$

where θ_1 and θ_2 are arbitrary constants which fix the boundary conditions of the Green's functions. (They are *not* related to the Bogoliubov transformations defined above.) We also find

$$H = \int d^3k\,\omega_k(A^\dagger_{2k}A_{1k} + A_{2k}A^\dagger_{1k}) = 2\int d^3k\,\omega_k\Gamma_{3k} \tag{42}$$

in the free-field case independent of θ_1 and θ_2.

The theory is not invariant under Bogoliubov transformations due to their noncommutativity with H. This is consonant with the absence of any evidence in nature for such an invariance (and related constants of motion). The point of our formulation is to ensure that representations of the field algebra and dynamics, which are related to each other by Bogoliubov transformations, are unitarily equivalent. In the case of flat space-time the unitary equivalence is a moot point, since a unique generator of the dynamics, the Hamiltonian, is apparent. In nonstatic space-times, where no unique generator of the dynamical motion is determined, the unitary equivalence is necessary in order to have an unambiguous quantum field theory (given the action).

Different choices for the generator of the dynamics lead to representations which can be related by Bogoliubov transformations. These representations are unitarily equivalent in our formulation, but not equivalent in the usual formulation. We return to this issue in sect. **4**.

The role of θ_1 and θ_2 is evident in the Green's functions. As usual we define the Green's functions as the vacuum expectation values of the time-ordered product of the field operators:

$$iG_{ij}(x-y) = {}_1\langle 0|T(\varphi_i(x)\varphi_j(y))|0\rangle_2 . \tag{43}$$

Equation (41) implies

$$G_{22}(x-y) = \sin^2\theta_2 G_F(x-y) + \cos^2\theta_2 G_F^*(x-y) , \tag{44}$$

where $G_F(x-y)$ is the usual Feynman propagator. G_{12} and G_{11} also depend on θ_1 and θ_2, but their precise expressions will not be of use in our presentation.

We shall now show that a model exists within our formulation which is physically equivalent to any conventional scalar quantum field theory with interaction $\mathscr{L}_I(\varphi)$. Our model Lagrangian is given by eq. (6) plus the interaction Lagrangian $\mathscr{L}_I(\varphi_2)$, where $\mathscr{L}_I(\varphi_2)$ is the same function of φ_2 as $\mathscr{L}_I(\varphi)$ is of φ. In order to have Feynman propagators, it is necessary to choose the boundary condition $\theta_2 = \pi/2$. We shall demonstrate that an asymptotic state exists in our formulation which corresponds to any asymptotic state of the usual formulation, and then show that S-matrix elements between corresponding states in the two models are equal in any order of perturbation theory.

The construction of asymptotic fields and states in our model is based on the renormalized quadratic part of the Lagrangian (eq. (6)). Therefore the previous development of this section can be used if appropriate subscripts « in » or « out » are appended to the operators. In particular, since $\theta_2 = \pi/2$, we have

$$\varphi_{2\text{in}}(x) = \int d^3k\,[f_k(x)A_{1k\text{in}} + f_k^*(x)A^\dagger_{2k\text{in}}] \tag{45a}$$

by eq. (41). We shall express the in-field operator of the usual formulation by

$$\varphi_{\text{in}}(x) = \int d^3k\,[f_k(x)A_{k\text{in}} + f_k^*(x)A^\dagger_{k\text{in}}] . \tag{45b}$$

Note that the form of $\varphi_{2\text{in}}$ and φ_{in} is identical except for the subscripts « 1 » and « 2 » on the operators. In addition, the commutation relations of the field operators and the Fourier-coefficient operators are also identical except for numerical subscripts. Furthermore, the application of the field operators to the vacua is also identical in effect (except for subscripts)

$$\begin{cases} \varphi_{\text{in}}(x)|0\rangle = \varphi^{(-)}_{\text{in}}(x)|0\rangle , & \varphi_{2\text{in}}(x)|0\rangle_2 = \varphi^{(-)}_{2\text{in}}(x)|0\rangle_2 , \\ \langle 0|\varphi_{\text{in}}(x) = \langle 0|\varphi^{(+)}_{\text{in}}(x) , & {}_1\langle 0|\varphi_{2\text{in}}(x) = {}_1\langle 0|\varphi^{(+)}_{2\text{in}}(x) , \end{cases} \tag{45c}$$

where the superscript « + » labels positive-frequency parts of the field operator and « − » labels negative-frequency parts. This close parallel in properties between our model and the model of the usual formulation implies the identity

$$\langle 0|\mathscr{P}(\varphi_{\text{in}})|0\rangle = {}_1\langle 0|\mathscr{P}(\varphi_{2\text{in}})|0\rangle_2 , \tag{45d}$$

where $\mathscr{P}(\varphi_{in})$ is any polynomial in the field φ_{in}. Later we shall use this identity to demonstrate the equality of the S-matrices in our model and the given model of the usual formulation. (Note that a straightforward application of eq. (45*d*) implies that the propagator G_{22} in our formulation equals the time-ordered propagator of the usual formulation.)

We now state the rule associating asymptotic particle states in our formulation with those of the usual formulation: given an in or out ket of the usual formulation, the corresponding ket in our formulation is obtained by appending the subscript « 2 » to every Fourier-coefficient operator (and to the vacuum) (*e.g.* $A^{\dagger}_{k in}|0\rangle \Leftrightarrow A^{\dagger}_{2k in}|0\rangle_2$). Given an in or out bra of the usual formulation, the corresponding bra in our formulation is obtained by appending « 1 » to each Fourier-coefficient operator (and to the vacuum) (*e.g.* $\langle 0|A_{k in} \Leftrightarrow$ $\Leftrightarrow {}_1\langle 0|A_{1k in}$). It is easily seen that energy-momentum eigenstates in the usual formulation correspond to energy-momentum eigenstates in our formulation. Thus we have identified the set of physical states in our model and find a detailed correspondence to those of the usual formulation.

The development of the perturbation theory of our model is completely analogous to the usual development. The S-matrix relates in and out fields: $\varphi_{2in}(x) = S\varphi_{2out}(x)S^{-1}$. LSZ reduction formulae are derived in the same manner as in the usual formulation. We find the reduction formula for a particle from an in-state and from an out-state to be, respectively,

$$(46)\quad \begin{cases} \langle \beta\ \text{out}|\alpha p\ \text{in}\rangle = \\ \qquad = \langle \beta - p\ \text{out}|\alpha\ \text{in}\rangle + \dfrac{i}{\sqrt{Z}}\displaystyle\int d^4x\, f_p(x)(\overrightarrow{\Box + m^2})\langle \beta\ \text{out}|\varphi_2(x)|\alpha\ \text{in}\rangle\,, \\ \langle \beta k\ \text{out}|\alpha\ \text{in}\rangle = \\ \qquad = \langle \beta\ \text{out}|\alpha - k\ \text{in}\rangle + \dfrac{i}{\sqrt{Z}}\displaystyle\int d^4x\, f_k^*(x)(\overrightarrow{\Box + m^2})\langle \beta\ \text{out}|\varphi_2(x)|\alpha\ \text{in}\rangle\,, \end{cases}$$

where $\varphi_2(x)$ is the interacting field and where we use the notation of ref. (9). The reduction of several particles leads to expressions which are identical to corresponding expressions of the usual model if the subscript « 2 » is appended to $\varphi(x)$.

Just as in the conventional model, we can formally develop a perturbation theory based on the U-matrix. The U-matrix relates the interacting and asymptotic field operator

$$(47a)\qquad \varphi_2(\boldsymbol{x}, t) = U^{-1}(t)\varphi_{2in}(\boldsymbol{x}, t)U(t)$$

---

(9) We follow the conventions and notation of J. D. BJORKEN and S. D. DRELL: *Relativistic Quantum Fields* (New York, N. Y., 1965).

and is easily shown to satisfy the differential equation

$$i\frac{\partial U}{\partial t} = -\left[\int \mathrm{d}^3x\, \mathscr{L}_1(\varphi_{2\mathrm{in}})\right] U\,. \tag{47b}$$

Defining $U(t,t') = U(t)U^{-1}(t')$ and solving eq. (47*b*) gives

$$U(t,t') = T\exp\left[i\int_{t'}^{t} \mathrm{d}^4x\, \mathscr{L}_1(\varphi_{2\mathrm{in}})\right]. \tag{48}$$

The LSZ procedure defined above reduces the calculation of S-matrix elements to the evaluation of time-ordered products of the interacting fields, ${}_1\langle 0|T(\varphi_2(x_1)\varphi_2(x_2)\dots\varphi_2(x_N))|0\rangle_2$. The U-matrix can then be used to reduce this quantity to the ratio of matrix elements involving only in-fields:

$$\frac{{}_1\langle 0|T(\varphi_{2\mathrm{in}}(x_1)\dots\varphi_{2\mathrm{in}}(x_N)\exp[i\int \mathrm{d}^4x\, \mathscr{L}_1(\varphi_{2\mathrm{in}})])|0\rangle_2}{{}_1\langle 0|T(\exp[i\int \mathrm{d}^4x\, \mathscr{L}_1(\varphi_{2\mathrm{in}})])|0\rangle_2}\,. \tag{49}$$

Expanding to any order in the interaction in eq. (49) gives matrix elements of polynomials in $\varphi_{2\mathrm{in}}$ which are equal—term by term—to corresponding matrix elements of the perturbation theory of the model of the conventional formulation by eq. (45*d*). Thus S-matrix elements between corresponding states are identically equal in the conventional model and our corresponding model.

It should be noted that only a subset of the possible asymptotic states in our model are identified as physical particle states which correspond to states in the usual model. The operator $A_{2k\mathrm{in}}$ can also be used to create in-kets (and $A^\dagger_{1k\mathrm{out}}$ to create out bras), but the S-matrix elements between physical kets and any ket (or bra) in which these operators appear is zero. (This follows from the fact that $[\mathscr{L}_1(\varphi_{2\mathrm{in}}), A_{2k\mathrm{in}}] = [\mathscr{L}_1(\varphi_{2\mathrm{in}}), A^\dagger_{1k\mathrm{in}}] = 0$ and ${}_1\langle 0|A_{2k\mathrm{in}} = 0 = A^\dagger_{1k\mathrm{in}}|0\rangle_2$.) Thus the S-matrix is block diagonal in our model. The part of it corresponding to the physical state sector is identical to the S-matrix of the given model of the conventional formulation.

The expression for the vacuum expectation value from which S-matrix elements may be calculated, eq. (49), can be used to show the unitary equivalence of representations which are related by a Bogoliubov transformation. Suppose that we had not used the representation of eq. (45*a*), but instead the Bogoliubov-transformed representation

$$\begin{aligned}\varphi^{\mathrm{B}}_{2\mathrm{in}}(x) = \int \mathrm{d}^3k\,[f_k(x)(A_{1k\mathrm{in}}\cosh\lambda + A^\dagger_{1k\mathrm{in}}\sinh\lambda) +{}\\ + f^*_k(x)(A^\dagger_{2k\mathrm{in}}\cosh\lambda + A_{2k\mathrm{in}}\sinh\lambda)] = B_{0\lambda}\varphi_{2\mathrm{in}}(x)B^{-1}_{0\lambda}\,.\end{aligned} \tag{50}$$

The canonical nature of the transformation guarantees that the canonical commutation relations will be maintained. If we follow the development of the perturbation theory given by eqs. (45)-(49) with φ^{B}_{2in} replacing φ_{2in} and $\varphi^{B}_{2} = U^{-1}(t)\varphi^{B}_{2in}U(t)$ replacing φ_2, then we find that S-matrix elements are calculated from vacuum expectation values involving only φ^{B}_{2in} fields:

$$(51) \qquad \frac{{}_1\langle 0|T(\varphi^{B}_{2in}(x_1)\dots\varphi^{B}_{2in}(x_N)\exp[i\int d^4x\,\mathscr{L}_1(\varphi^{B}_{2in})])|0\rangle_2}{{}_1\langle 0|T(\exp[i\int d^4x\,\mathscr{L}_1(\varphi^{B}_{2in})])|0\rangle_2}.$$

Since $B^{-1}_{0\lambda}|0\rangle_2 = |0\rangle_2$ and ${}_1\langle 0|B_{0\lambda} = {}_1\langle 0|$ we find that eq. (51) is equal to eq. (49). Thus the unitary equivalence of representations of the quantum field theory differing by a Bogoliubov transformation is demonstrated. (One can formally define Bogoliubov transformations for interacting fields $B^{int} = UBU^{-1}$, but B^{int} is not unitary due to the well-known difficulties of the U-matrix in the conventional formulation which are also present in our formulation. We circumvent this problem by working with the definition of the S-matrix in terms of vacuum expectation values of asymptotic in-fields, where the unitary equivalence under Bogoliubov transformation can be unambiguously shown to hold.)

A comparison of our formulation with the usual formulation shows a certain similarity of form at the Lagrangian level if our Lagrangian is put in the form

$$(52) \qquad \mathscr{L} = -\frac{1}{2}\partial_\mu(\varphi_1-\varphi_2)\partial^\mu(\varphi_1-\varphi_2) + \partial_\mu\varphi_2\partial^\mu\varphi_2 + \\ + \mathscr{L}_1(\varphi_2) + \frac{m^2}{2}(\varphi_1-\varphi_2)^2 - \frac{m^2}{2}\varphi_2^2.$$

In the usual approach $\varphi_3 = \varphi_1 - \varphi_2$ is an ignorable field and it would not have been surprising that we found equal S-matrix elements above. However, our formulation differs from the usual formulation in two respects—first, the field operators are both expanded in type «1» and «2» Fourier coefficient operators and, more importantly, the vacuum is defined in a way which correlates the φ_3 and φ_2 sectors. In the $\theta_2 = \pi/2$ case the first difference can be eliminated by a relabeling of Fourier-coefficient operators. However, for other values of θ_2 both differences are present and lead to a very different theory from the usual formulation. While it is clear that the correlation between the φ_3 and φ_2 sectors can be implemented in free field theory, one might ask if this remains true in the interacting case. Certainly the correlation can be implemented in the asymptotic fields and states, since that is free field theory. One can also *formally* implement the correlation for interacting fields through eq. (47*a*). But the implementation of the correlation in the interacting case is actually based on the reduction of the S-matrix element to the vacuum expectation value of products of asymptotic fields. Since the correlation can be maintained for the asymptotic fields and states, the physical quantities of the models, S-matrix elements, embody the correlation. (The value of the correlation we introduce is twofold: first, it is necessary in order to obtain the unitary equivalence of Bogoliubov rotated representations, and secondly, it widens the range of allowed flat-space-time quantum field theories to include those with principal-value propagators ([8]).)

We conclude this section with a brief discussion of our formulation of the charged-scalar-particle case. In the usual approach, the free-charged-scalar-particle Lagrangian may be expressed in terms of complex fields, $\varphi(x)$ and $\varphi^*(x)$ or in terms of two real fields $\varphi_a(x)$ and $\varphi_b(x)$ with

$$(53)\qquad \varphi(x) = [\varphi_a(x) + i\varphi_b(x)]/\sqrt{2}\,.$$

If we follow the same procedure as above for the real fields, double their number and use the Lagrangian form of eq. (6), we are led to the complex field expression of the Lagrangian:

$$(54)\qquad \mathscr{L} = \partial_\mu\tilde{\varphi}_2\partial^\mu\varphi_1 + \partial_\mu\varphi_2\partial^\mu\tilde{\varphi}_1 - \partial_\mu\tilde{\varphi}_1\partial^\mu\varphi_1 - m_2\tilde{\varphi}_2\varphi_1 - m^2\varphi_2\tilde{\varphi}_1 + m^2\tilde{\varphi}_1\varphi_1\,,$$

where

$$(55)\qquad \varphi_i(x) = [\varphi_{ia}(x) + i\varphi_{ib}(x)]/\sqrt{2}$$

and

$$(56)\qquad \tilde{\varphi}_i(x) = [\varphi_{ia}(x) - i\varphi_{ib}(x)]/\sqrt{2}\,.$$

We require φ_{ia} and φ_{ib} to embody the same boundary conditions, so that the expansion of φ_{ia} and φ_{ib} utilizes the same constants, c_{ij} and $\tilde{c}_{ij}$. Consequently

$$(57)\qquad \varphi_i(x) = \int d^3k[(c_{i1}A_{+1k} + c_{12}A_{+2k})f_k + (\tilde{c}_{i1}A^\dagger_{-1k} + \tilde{c}_{i2}A^\dagger_{-2k})f^*_k]\,,$$

$$(58)\qquad \tilde{\varphi}_i(x) = \int d^3k[(c_{i1}A_{-1k} + c_{i2}A_{-2k})f_k + (\tilde{c}_{i1}A^\dagger_{+1k} + \tilde{c}_{i2}A^\dagger_{+2k})f^*_k]\,,$$

where φ_i and $\tilde{\varphi}_i$ are related via the charge conjugation operator ([9]). Following the quantization pattern discussed above, with only minor changes due to the presence of two types of Fourier coefficient operators: positive charge, A_{+ik}, and negative charge, A_{-ik}, leads eventually to the following Green's function:

$$(59)\qquad G_{22}(x-y) = G_F(x-y)\sin^2\theta_2 + G^*_F(x-y)\cos^2\theta_2\,.$$

Note that it has the same form as eq. (1). (Lagrangian interaction terms are expressed solely in terms of φ_2 and $\tilde{\varphi}_2$.) As a result we require $\gamma\tilde{\varphi}_2\gamma^{-1} = \varphi_2^\dagger$, the Hermitian conjugate of φ_2, where γ is the metric operator so that

$$(60)\qquad C\varphi_2C^{-1} = \gamma\varphi_2^\dagger\gamma^{-1}\,,$$

where C is the charge conjugation operator.

3. – Fermion quantization.

In this section we describe our formulation of spin–one-half fermion quantum field theory. Again we are motivated by the need for a unique particle interpretation in nonstatic space-time. The formulation for fermions has close similarities to boson quantization.

Two fields are needed to describe a spin–one-half particle. The Lagrangian is

$$\mathscr{L} = \tilde{\psi}_2\gamma^0(i\nabla - m)\psi_1 + \tilde{\psi}_1\gamma^0(i\nabla - m)\psi_2 - \tilde{\psi}_1\gamma^0(i\nabla - m)\psi_1 . \tag{61}$$

We follow the conventions and notation of ref. (9). The fields $\tilde{\psi}_i$ will be related to the transpose of the charge conjugate field via

$$\tilde{\psi}_i = \psi_i^{cT}\gamma^0 C^T \tag{62}$$

for $i = 1, 2$. (In the usual formulation $\psi^\dagger = \tilde{\psi}$ would hold.) The equations of motion are

$$(i\nabla - m)\psi_i = 0 , \qquad \tilde{\psi}_i(i\overleftarrow{\nabla} - m) = 0 , \tag{63}$$

for $i = 1, 2$. The momentum conjugate to ψ_1 is

$$\pi_1 = i(\tilde{\psi}_2 - \tilde{\psi}_1) \tag{64}$$

and the conjugate to ψ_2 is

$$\pi_2 = i\tilde{\psi}_1 . \tag{65}$$

The canonical equal-time anticommutation relations imply

$$\{\psi_{1\alpha}(x), \tilde{\psi}_{1\beta}(y)\} = 0 , \tag{66}$$

$$\{\psi_{1\alpha}(x), \tilde{\psi}_{2\beta}(y)\} = \delta_{\alpha\beta}\delta^3(\boldsymbol{x} - \boldsymbol{y}) \tag{67}$$

and

$$\{\psi_{2\alpha}(x), \tilde{\psi}_{2\beta}(y)\} = \delta_{\alpha\beta}\delta^3(\boldsymbol{x} - \boldsymbol{y}) . \tag{68}$$

The most general form for the mode of expansion of the fields is (9)

$$\psi_i = \sqrt{2m}\sum_s \int \mathrm{d}^3k\,[(c_{i1}b_{1ks} + c_{i2}b_{2ks})f_k(x)u_{ks} + (\tilde{c}_{i1}d^\dagger_{1ks} + \tilde{c}_{i2}d^\dagger_{2ks})f^*_k(x)v_{ks}] . \tag{69}$$

Just as in the charged scalar case, we develop our formulation in such a way that the even and odd charge conjugation combinations, $\psi_i \pm \psi_i^c$, implement the same boundary conditions. Therefore

$$(70)\qquad \tilde{\psi}_i = \sqrt{2m}\sum_s \int d^3k[(c_{i1}d_{1ks} + c_{i2}d_{2ks})f_k(x)v_{ks}^\dagger + (\tilde{c}_{i1}b_{1ks}^\dagger + \tilde{c}_{i2}b_{2ks}^\dagger)f_k^*(x)u_{ks}^\dagger].$$

The nonzero Fourier-coefficient anti-commutation relations are

$$(71)\qquad \{d_{iks}, d_{jk's'}^\dagger\} = \{b_{iks}, b_{jk's'}^\dagger\} = (1-\delta_{ij})\delta_{ss'}\delta^3(\boldsymbol{k}-\boldsymbol{k}')$$

for $i, j = 1, 2$. The definition of states and inner products mirror the boson case. The vacua are defined by

$$(72)\qquad b_{1ks}|0\rangle_2 = b_{1ks}^\dagger|0\rangle_2 = d_{1ks}|0\rangle_2 = d_{1ks}^\dagger|0\rangle_2 = 0$$

and

$$(73)\qquad b_{2ks}|0\rangle_1 = b_{2ks}^\dagger|0\rangle_1 = d_{2ks}|0\rangle_1 = d_{2ks}^\dagger|0\rangle_1 = 0$$

and are related by a metric operator η:

$$(74)\qquad \eta|0\rangle_1 = |0\rangle_2$$

which satisfies $\eta = \eta^\dagger = \eta^{-1}$. We conventionally choose to construct kets from $|0\rangle_2$ and define their dual as their Hermitian conjugate multiplied by the metric operator. Thus inner products have the form

$$(75)\qquad \langle\alpha|\beta\rangle = (|\alpha\rangle)^\dagger\eta|\beta\rangle .$$

Physical observables must be self-adjoint, $A = A^* = \eta^{-1}A^\dagger\eta$, in order to have real eigenvalues. The Hamiltonian must be self-adjoint in order to conserve the norm. In view of eq. (68) we only use ψ_2 and $\tilde{\psi}_2$ in interaction terms and therefore require $\tilde{\psi}_2 = \eta^{-1}\psi_2^\dagger\eta$, so that

$$(76)\qquad C\psi_2C^{-1} = \eta^{-1}C\tilde{\psi}_2^T\eta ,$$

which bears comparison with eq. (15.112) of ref. (9) and also eq. (60) above. The equal-time anticommutation relations, eq. (76), and the adjointness of H restrict the constants c_{ij} and $\tilde{c}_{ij}$ so that

$$(77)\qquad \psi_1 = \sqrt{2m}\sum_s \int d^3k[(\cos(\theta_1-\theta_2)b_{1ks} + \sin(\theta_1-\theta_2)b_{2ks})f_k(x)u_{ks}/\sin\theta_1 + \\ + (\cos(\theta_1-\theta_2)d_{1ks}^\dagger - \sin(\theta_1-\theta_2)d_{2ks}^\dagger)f_k^*(x)v_{ks}/\cos\theta_1]$$

and

$$(78)\qquad \psi_2 = \sqrt{2m}\sum_s \int d^3k[(\sin\theta_2 b_{1ks} + \cos\theta_2 b_{2ks})f_k(x)u_{ks} + \\ + (\cos\theta_2 d_{1ks}^\dagger + \sin\theta_2 d_{2ks}^\dagger)f_k^*(x)v_{ks}] .$$

The Hamiltonian is

$$H = \sum_s \int d^3k\omega_k(b^\dagger_{2ks}b_{1ks} - b_{2ks}b^\dagger_{1ks} + d^\dagger_{2ks}d_{1ks} - d_{2ks}d^\dagger_{1ks}) . \tag{79}$$

In contrast to the usual formulation, we see that our Hamiltonian does not have an infinite vacuum energy with respect to $|0\rangle_2$. It is not positive definite, but we will be able to develop a unitary S-matrix theory in the space of positive-energy asymptotic states, if we choose $\theta_2 = \pi/2$. This is evident from an examination of the Green's function

$$S_{22}(x-y) = -i\,{}_1\langle 0|T(\psi_2(x)\tilde{\psi}_2(y)\gamma_0)|0\rangle_2 = \tag{80}$$

$$= \sin^2\theta_2 S_F(x-y) + \cos^2\theta_2 C\gamma^0 S^*_F(C\gamma^0)^{-1} , \tag{81}$$

which gives $S_{22} = S_F$, the usual Feynman propagator, if $\theta_2 = \pi/2$. As in the boson case, we introduce interactions only through type-2 operators, and use type-2 operators to LSZ reduce in and out particles. The result is a perturbation theory which, in the Fermion sector, only involves time-ordered products of ψ_2 and $\tilde{\psi}_2$. Thus we can establish a model in our formulation which is equivalent, so far as S-matrix elements are concerned, to any given model of the usual formulation. In particular, our model electrodynamics has the Lagrangian

$$\mathscr{L} = \tfrac{1}{2}F^1_{\mu\nu}F^{2\mu\nu} + \tfrac{1}{4}F^1_{\mu\nu}F^{1\mu\nu} + \tilde{\psi}_2\gamma^0(i\nabla - m)\psi_1 + \tilde{\psi}_1\gamma^0(i\nabla - m)\psi_2 - \tilde{\psi}_1\gamma^0(i\nabla - m)\psi_1 - e_0\tilde{\psi}_2\gamma^0\hat{A}_2\psi_2 \tag{82}$$

with

$$F^i_{\mu\nu} = \partial_\nu A_{i\mu} - \partial_\mu A_{i\nu} \tag{83}$$

for $i = 1, 2$. While we will discuss this model more fully elsewhere [6] two things are worth noting. First the interaction is expressed solely in terms of fields of type 2—both for fermions and the photon. Following our quantization procedure leads to S-matrix expressions which are term-by-term equal to corresponding expressions in QED. Secondly, the model is gauge invariant. The gauge transformation is

$$\psi_2 \to \exp[i\Lambda]\psi_2 , \tag{84}$$

$$\tilde{\psi}_2 \to \exp[-i\Lambda]\tilde{\psi}_2 , \tag{85}$$

$$A_{2\mu} \to A_{2\mu} - \frac{1}{e}\partial_\mu\Lambda , \tag{86}$$

$$\psi_1 \to \psi_1 + (\exp[i\Lambda] - 1)\psi_2 , \tag{87}$$

$$\tilde{\psi}_1 \to \tilde{\psi}_1 + (\exp[-i\Lambda] - 1)\tilde{\psi}_2 , \tag{88}$$

and its associated conserved current is

$$J_\mu = -i\frac{\delta \mathscr{L}}{\delta \partial^\mu \psi_1}\psi_2 - i\frac{\delta \mathscr{L}}{\delta \partial^\mu \psi_2}\psi_2 , \tag{89}$$

$$J_\mu = \bar{\psi}_2 \gamma^0 \gamma_\mu \psi_2 . \tag{90}$$

We close our discussion of fermions by considering the case $\theta_2 = \pi/4$ which, by eq. (81), gives the principal-value propagator

$$S_{22}(x-y) = \int \frac{d^4k}{(2\pi)^4} \exp[-ik\cdot(x-y)](\not{k}+m)\frac{P}{k^2-m^2}, \tag{91}$$

where

$$\frac{P}{k^2-m^2} = \frac{1}{2}\left(\frac{1}{k^2-m^2+i\varepsilon} + \frac{1}{k^2-m^2-i\varepsilon}\right). \tag{92}$$

Thus a quantum action-at-a-distance model of fermions can be constructed within the framework of our formulation.

4. – Particle interpretation.

In this section we shall show that the particle interpretation of our formulation of quantum field theory is well defined for the case of a free scalar particle in a nonstatic space-time where no global timelike co-ordinate exists. We assume an action of the form

$$S = \int d^4x\,[\varphi_2 D\varphi_1 - \tfrac{1}{2}\varphi_1 D\varphi_1] + \text{(surface terms)}, \tag{93}$$

which under the variation of S gives the field equations

$$D\varphi_1 = D\varphi_2 = 0 . \tag{94}$$

The self-adjointness of D implies

$$\int_V [f^* Dg - (Df)^* g]\, d^4x = \int_{\Sigma_v} f^* \overleftrightarrow{D}^\mu g\, d\Sigma_\mu , \tag{95}$$

where Σ_v is the surface bounding V, $d\Sigma_\mu$ is an outward directed surface element of Σ_v and $\overleftrightarrow{D}^\mu$ is a two-edged vector differential operator. If Σ is a spacelike complete Cauchy hypersurface for the field equations (we assume they exist), then an inner product for complex solutions of the field equations can be defined by

$$(v_1, v_2) = i\int_\Sigma v_1^* \overleftrightarrow{D}^\mu v_2 \,\mathrm{d}\Sigma_\mu . \tag{96}$$

We now choose an arbitrary complete orthonormal set of pairs of complex conjugate solutions of eq. (94), $\{V_\alpha, V_\alpha^*\}$, satisfying

$$(V_\alpha, V_{\alpha'}) = -(V_\alpha^*, V_{\alpha'}^*) = \delta_{\alpha\alpha'} , \tag{97}$$

$$(V_\alpha, V_{\alpha'}^*) = 0 , \tag{98}$$

and use them in the mode expansion of the field operators

$$\varphi_i = \sum_\alpha [(c_{i1}A_{1\alpha} + c_{i2}A_{2\alpha}) V_\alpha + (\tilde{c}_{i1}A_{1\alpha}^\dagger + \tilde{c}_{i2}A_{2\alpha}^\dagger) V_\alpha^*] , \tag{99}$$

where c_{ij} and $\tilde{c}_{ij}$ are real c-numbers. The Fourier coefficient operators satisfy

$$[A_{i\alpha}, A_{j\alpha'}^\dagger] = (1-\delta_{ij})\delta_{\alpha\alpha'} \tag{100}$$

with all other commutators equal to zero. The commutativity of Fourier-coefficient operators and their Hermitian conjugate allows us to define the vacua $|0\rangle_1$ and $|0\rangle_2$ by

$$A_{1\alpha}|0\rangle_2 = A_{1\alpha}^\dagger|0\rangle_2 = A_{2\alpha}|0\rangle_1 = A_{2\alpha}^\dagger|0\rangle_1 = 0 . \tag{101}$$

We choose to construct states from $|0\rangle_2$. The one-particle ket corresponding to the Fourier transform variable α is

$$|\alpha\rangle = -(v_\alpha^*, \varphi_2)|0\rangle_2/\tilde{c}_{22} \tag{102}$$

and the one-particle bra dual to it is

$$\langle\alpha| = (|\alpha\rangle)^\dagger \gamma , \tag{103}$$

where γ is a metric operator satisfying $\gamma = \gamma^{-1} = \gamma^\dagger$ and

$$\gamma^{-1} A_{i\alpha}\gamma = \varepsilon_{ij}A_{j\alpha} \tag{104}$$

with $\varepsilon_{11} = \varepsilon_{22} = 0$ and $\varepsilon_{12} = \varepsilon_{21} = 1$. The further development of this quantization proceeds along the lines of sect. **2**. In particular φ_2 is self-adjoint.

We now introduce a quantization of the particle described by S which parallels the above development in every detail except that a different complete orthonormal set of field equation solutions $\{W_\beta, W_\beta^*\}$ is used in the mode expansion of the fields

$$\varphi_i = \sum_\beta [(c_{i1} A_{1\beta} + c_{i2} A_{2\beta}) W_\beta + (\tilde{c}_{i1} A^\dagger_{1\beta} + \tilde{c}_{i2} A^\dagger_{2\beta}) W^*_\beta]. \tag{105}$$

The question arises: how are we to relate the two quantizations? In the usual formulation only one answer is apparent—the field operators are to be identified [1,2], since they are uniquely determined by the field equations and the canonical commutation relations. But in the present case, the field operators are not uniquely determined, so that the identification of the fields in eq. (105) and (99) is not required. The relation between the quantizations must obviously be well defined (in the sense that every operator and state in one quantization can be uniquely expressed in terms of operators and states of the other representations). More importantly, it must only relate operators whose properties are fixed by the field equation and the canonical commutation relations; and whose expectation values are uniquely specified by purely geometrical restrictions on their support and do not embody a definition of positive frequency. In our formalism, the operators which satisfy these requirements are linear combinations of

$$\varphi^{\text{II}}_{iv} = \sum_\alpha (A_{i\alpha} V_\alpha + A^\dagger_{i\alpha} V^*_\alpha), \tag{106}$$

$$\varphi^{\text{II}}_{iw} = \sum_\beta (A_{i\beta} W_\beta + A^\dagger_{i\beta} W^*_\beta), \tag{107}$$

for $i = 1, 2$ in the respective quantizations we can restrict the discussion to these quantities. In particular, the vacuum expectation value

$${}_1\langle 0|\varphi^{\text{II}}_1(x)\varphi^{\text{II}}_2(y)|0\rangle_2 = \tfrac{1}{2}\,{}_1\langle 0|[\varphi^{\text{II}}_1(x), \varphi^{\text{II}}_2(y)]|0\rangle_2, \tag{108}$$

$${}_1\langle 0|\varphi^{\text{II}}_1(x)\varphi^{\text{II}}_2(y)|0\rangle_2 = \frac{i}{2}\Delta(x-y), \tag{109}$$

where $\Delta(x-y)$, the commutator function, has the well-defined geometrical property that it vanishes at spacelike distances.

Identifying φ^{II}_{iw} with φ^{II}_{iv}, for $i = 1, 2$, leads to the relations

$$A_{i\beta} = (W_\beta, \varphi^{\text{II}}_i) - \sum_\alpha [(W_\beta, V_\alpha) A_{i\alpha} + (W_\beta, V^*_\alpha) A^\dagger_{i\alpha}], \tag{110}$$

for $i = 1, 2$ plus Hermitian-conjugate expressions. The form of the inner products on the right-hand side of eq. (110) is determined by requiring that the definition of positive frequency implicit in the separation of the orthonormal set $\{W_\beta, W^*_\beta\}$ into complex conjugate pairs of solutions can also be implemented by linear combinations of V_α and V^*_α. Specifically, we assume that a complete orthonormal set of pairs of complex conjugate functions, $\{V_\alpha, V^*_\alpha\}$, exists which satisfies

$$V_\alpha = c_1 \tilde{V}_\alpha + c_2 \tilde{V}^*_\alpha , \tag{111}$$

$$V^*_\alpha = c^*_1 \tilde{V}^*_\alpha + c^*_2 \tilde{V}_\alpha , \tag{112}$$

for all α, where c_1 and c_2 are c-number functions of α only with $|c_1| > |c_2|$, and where

$$(W_\beta, \tilde{V}^*_\alpha) = 0 , \tag{113}$$

$$(W^*_\beta, \tilde{V}_\alpha) = 0 , \tag{114}$$

for all β. The orthogonality conditions imply

$$c_1 = \exp[i\lambda_1] \cosh \lambda_2 , \tag{115}$$

$$c_2 = \exp[i\lambda_1] \sinh \lambda_2 , \tag{116}$$

where λ_1 and λ_2 are solely functions of α. The substitution of eqs. (111) and (112) in eq. (110) and use of eqs. (113) and (114) gives

$$A_{i\beta} = \sum_\alpha (W_\beta, \tilde{V}_\alpha)[\exp[i\lambda_1] \cosh \lambda_2 A_{i\alpha} + \exp[-i\lambda_1] \sinh \lambda_2 A^\dagger_{i\alpha}] \tag{117}$$

for $i = 1, 2$. Note that the bracketed term on the right-hand side of the equation has the same form as the Bogoliubov rotated Fourier-coefficient operator given in eq. (23). In the present case we can rewrite eq. (117) in the form

$$A_{i\beta} = \sum_\alpha (W_\beta, \tilde{V}_\alpha) B_{\lambda_1\lambda_2} A_{i\alpha} B^{-1}_{\lambda_1\lambda_2} \tag{118}$$

with

$$B_{\lambda_1\lambda_2} = \exp\left[2i \sum_\alpha \lambda_1(\alpha) \Gamma_{3\alpha}\right] \exp\left[2i \sum_\alpha \lambda_2(\alpha) \Gamma_{2\alpha}\right] , \tag{119}$$

where $\Gamma_{3\alpha}$ and $\Gamma_{2\alpha}$ are obtained from eqs. (25) and (26) by replacing the subscripts k with α.

The particle interpretations of the two quantizations will now be shown to be identical. First we note that the vacuum $|0\rangle_2$ of the « α » quantization is invariant under $B_{\lambda_1\lambda_2}$, so that it may be taken to be identical with the $|0\rangle_2$ vacumm of the « β » quantization. Next we note that the canonical commutation relations and the vacuum expectation value of any product of field operators are invariant under B:

$${}_1\langle 0|\varphi_{i_1}(x_1)\varphi_{i_2}(x_2) \dots |0\rangle_2 = {}_1\langle 0|B^{-1}\varphi_{i_1}(x_1)\varphi_{i_2}(x_2) \dots B|0\rangle_2 . \tag{120}$$

This implies that we could replace $A_{i\alpha}$ with $B_{\lambda_1\lambda_2}A_{i\alpha}B^{-1}_{\lambda_1\lambda_2}$ in the mode expansions, eq. (107), with no change in physical consequences. In particular, this applies to the definition of particle kets. Equation (102) becomes

$$|\alpha\rangle = (\exp[-i\lambda_1]\cosh\lambda_2 A^\dagger_{2\alpha} + \exp[i\lambda_1]\sinh\lambda_2 A_{2\alpha})|0\rangle_2 . \tag{121}$$

Consequently, $A^\dagger_{2\beta}|0\rangle_2$ is a superposition of one-particle states in the « α » quantization. In general, the N-particle state in the « β » quantization is a superposition of N-particle states in the « α » quantization.

The invariance of particle number under Bogoliubov transformations is reflected in the relation between the particle number operator,

$$N = \sum_\alpha (A^\dagger_{2\alpha}A_{1\alpha} - A_{2\alpha}A^\dagger_{1\alpha}) , \tag{122}$$

which is invariant under Bogoliubov transformations, and related to the Casimir operator of the Bogoliubov group (cf. eqs. (29)-(31)). Our identification of N as the particle number operator is based, as most charge and number operators are, on an invariance of the action under a global change of phase of fields. In our case we note that the action of eq. (93) is invariant under the infinitesimal phase change

$$\varphi_1 \to \varphi_1 + i\varepsilon\varphi_1 , \qquad \varphi_2 \to \varphi_2 + i\varepsilon(\varphi_1 - \varphi_2) . \tag{123}$$

The corresponding conserved-number operator is given by

$$N = i\int_{\Sigma_\nu} \varphi_1 \overleftrightarrow{D}^\mu \varphi_2 \, d\Sigma_\mu . \tag{124}$$

Because φ_1 and φ_2 implicitly embody a definition of positive frequency, we are led to replace them with Hermitian operators:

$$N = i\int_{\Sigma_\nu} \varphi^{\text{H}}_1 D^\mu \varphi^{\text{H}}_2 \, d\Sigma_\mu \tag{125}$$

with φ^{H}_i given in eq. (106). Equation (125) can be evaluated by using eq. (96) and (97) to give eq. (122). Thus our definition of number operator is physically motivated. It is also consistent with our expectations of a number operator.

We shall now summarize our picture of second quantization in curved space-time where no global timelike Killing vector is present. Consider a complete spacelike Cauchy hypersurface. At each point on the surface there is a local timelike direction. There is, in general, a class of operators, which will locally generate a displacement in the timelike direction, but which globally generate very different motions. Due to the absence of a global timelike Killing vector, no member of the class of potential generators of the dynamics is physically selected as the generator of the dynamics. One is free to choose any member as the generator of the dynamics locally. Each choice implies a different definition of positive and negative frequency when field operators are represented by Fourier expansions.

In the usual formulation of quantum field theory any choice of generator of the dynamics (and thus Fourier representation of field operators) is unitarily inequivalent to any other choice in general. As a result each choice gives a *different physical theory* and the second quantization of a theory is not unique. Practically, this means that 1) a one-particle state in one quantization is a many-particle state in any other quantization (particle number is ambiguous), 2) in general one can construct a one-particle state which is an eigenstate of a generator in one representation, but one cannot construct a one-particle state in another representation which is an eigenstate of the same generator (the space of states is different), and 3) (if interactions are introduced) the S-matrix differs from quantization to quantization. Obviously, there are only two acceptable alternatives in this situation; either some new principle selects one representation as the correct physical representation, or a modification of quantum field theory is necessary. In the absence of a new physical principle, we have formulated a modification of quantum field theory.

Our formulation allows one to quantize a field theory with any of the potential generators of the dynamics and yet to have a physically unique theory. Different quantizations can be related by Bogoliubov transformations and, in our formulation, are unitarily equivalent. Consequently, particle number is invariant—N-particle states in one representation are superpositions of N-particle states in any other representation; the set of states in one representation is unitarily equivalent to the set of states in any other representation; and the S-matrix is uniquely determined in the case of an interacting theory (the proof is analogous to that of the flat space-time case discussed in sect. 2). Our formulation associates a unique physical theory with any given action. In a sense, it implements an equivalence principle in the space of solutions to the field equations—any complete orthonormal set of solutions to the field equations can be used in the Fourier expansion of field operators and a unique physical theory results (cf. eqs. (111)-(114)).

In conclusion, we note that the problem we have addressed relates to one observer and the ambiguities of conventional quantum field theory he must face. Different observers in relatively accelerating frames will not see the same number of particles in our formulation. Neither is particle creation near black holes precluded in our formulation.

* * *

I am grateful to the Aspen Center for Physics for its hospitality while part of this work was being done, and to M. A. B. Beg, S. Mandelstam and D. Park for stimulating conversations.

Appendix B. New Framework for Gauge Field Theories

This appendix reprints S. Blaha, "New Framework for Gauge Field Theories", IL Nuovo Cimento **49A**, 113 (1979).[156] It describes the PseudoQuantization of gauge field theories for the purposes of defining higher derivative field theories and for use in the quantization of fields in universes and the Megaverse in chapter 7.

[156] © Copyright Stephen Blaha 1978.

IL NUOVO CIMENTO Vol. 49 A, N. 1 1 Gennaio 1979

New Framework for Gauge Field Theories (*).

S. Blaha

Physics Department, Williams College - Williamstown, Ma. 01267

(ricevuto l'8 Agosto 1978)

Summary. — We formulate gauge theories within the framework of a generalization of quantum field theory. In particular, we discuss models of electrodynamics and of Yang-Mills theories, a model of the strong interaction with higher-order derivatives and quark confinement and a renormalizable model of pure quantum gravity with Einstein Lagrangian. In the case of electrodynamics we show that two models are possible: one with predictions which are identical to QED and one which is a quantum action-at-a-distance model of electrodynamics. In the case of Yang-Mills theories we can construct a model which is identical in predictions to any conventional model, or a quantum action-at-a-distance model. In the second case it is possible to eliminate all loops of Yang-Mills particles (in all gauges) in a manner consistent with unitarity. A variation of Yang-Mills models exists in our formulation which has higher-order derivative field equations. It is unitary and has positive probabilities. It can be used to construct a model of the strong interactions which has a linear potential and manifest quark confinement. Finally we show how to construct an action-at-a-distance model of pure quantum gravity (whose classical limit is the dynamics of the Einstein Lagrangian) coupled to an external classical source. The model is trivially renormalizable.

1. – Introduction.

Because of the absence of an acceptable physical interpretation of conventional quantum field theory in the case of curved nonstatic space-time, we

(*) Supported in part by Research Corporation and the National Science Foundation.

recently developed a modified formulation of quantum field theory (¹). We showed it had a unique physical particle interpretation in nonstatic space-time. We described the flat–space-time formulation for the cases of scalar particles and spin–one-half fermions. In both cases we found that it was possible to construct a model which was equivalent in predictions to any model of the usual formulation. However, it was also possible to construct models which had no analogue in the usual formulation. In these models the quantum exchange of a « particle » did not take place via a Feynman propagator. Rather, the propagator G could be chosen to be any of a continuum of possibilities ranging from the Feynman propagator to the principal-value (half advanced-half retarded) propagator. The choice is parametrized by an angle θ, whose specification is equivalent to a choice of boundary condition

$$G(x-y) = \sin^2\theta\, G_F(x-y) + \cos^2\theta\, C G_F^* C^{-1}\,, \tag{1}$$

where G_F is the usual Feynman propagator with G_F^* its complex conjugate and C is an appropriate charge conjugation matrix.

The new degree of freedom, represented by θ, leads to new possibilities for the formulation of flat–space-time quantum field theory models which will be especially evident in the case of gauge theories.

In sect. **2** we shall explore models of quantum electrodynamics which occur within the framework of our formulation. We shall show that one model is completely equivalent to QED in its predictions. In addition, we shall show that a quantum action-at-a-distance electrodynamics is also possible.

In sect. **3** we describe model Yang-Mills theories and show that a model equivalent to any conventional model can be formulated as well as a quantum action-at-a-distance model. In the second case all non-Abelian boson loops can be eliminated (in all gauges) in a manner which is consistent with unitarity.

In sect. **4** we describe a non-Abelian model of quark confinement based on higher-derivative field equations for which a unitary physical S-matrix is obtained by the choice of principal-value propagators. This demonstrates the utility of principal-value propagators for widening the class of unitary quantum field theories to include higher-derivative theories.

In sect. **5** we describe a model of pure quantum gravity based on the Einstein Lagrangian which is trivially renormalizable, if gravitons propagate via principal-value propagators. (No graviton loops.) This illustrates the potential of the principal-value propagator to ameliorate renormalization problems by allowing one to limit the self-interactions of quantum fields.

In view of the unfamiliar features of principal-value propagators, we shall

(¹) S. BLAHA: *Nuovo Cimento*, **49** A, 35 (1978).

devote the remainder of this section to a discussion of their properties in relation to causality and analyticity.

Among the first appearances of principal-value, or half advanced-half retarded, propagators was in Feynman's space-time approach to quantum electrodynamics ([2]), where he observed that one is naturally led to such a propagator for the photon, if one considers a quantum one-photon exchange between two electrons wherein the photons propagate, as in classical electrodynamics, via retarded propagators. Since one cannot distinguish between the process where a photon propagates from electron A to B and the process where the photon propagates from B to A for short time intervals one must sum the two amplitudes and a principal-value propagator results. This observation illustrates the absence of a direct relation between the propagator of the classical field and the propagator of the corresponding quantum field. In particular, it is perfectly possible for a quantum field to have a principal-value propagator, and yet have the classical field propagate via a retarded propagator. This can be understood within the framework of the absorber model of Feynman and Wheeler ([3]). For a quantum process, where a finite number of quanta is exchanged, the effective propagator of the quanta cannot have this aspect of its character changed. But in a classical situation, where infinite numbers of quanta are involved, the effective propagator of the quanta can be changed through interaction with the absorber in the manner outlined in ref. ([3]). We therefore conclude that the propagation of quanta via principal-value propagators does not imply that macroscopic causality (as represented by retarded propagators) is lost. This remark is relevant to our models of the strong interaction and quantum gravity discussed later.

So far as microscopic causality is concerned, it will be seen that principal-value propagators are completely consistent with the vanishing of commutators of field operators for spacelike distances. Thus neither macroscopic nor microscopic causality is necessarily inconsistent with the use of quantum principal-value propagators.

The use of principal-value propagators does lead to a different analytic structure of S-matrix amplitudes. Amplitudes are now piecewise analytic. However, several authors ([4]) have shown that such amplitudes are not necessarily inconsistent with experiment—even for the case of the pion-nucleon dispersion relations. In particular, considering the absence of any direct relation between the analytic properties of quark-quark scattering amplitudes,

---

([2]) R. Feynman: *Phys. Rev.* **76**, 769 (1949), footnote ([7]).
([3]) J. Wheeler and R. Feynman: *Rev. Mod. Phys.*, **17**, 157 (1945); **21**, 425 (1949); F. Hoyle and J. Narlikar: *Ann. of Phys.*, **54**, 207 (1969); **62**, 44 (1971).
([4]) R. E. Cutkowsky, P. V. Landshoff, D. I. Olive and J. C. Polkinghorne: *Nucl. Phys.*, **12** B, 281 (1969); T. D. Lee and G. C. Wick: *Phys. Rev. D*, **2**, 1033 (1970); M. Gundzik and E. C. G. Sudarshan: *Phys. Rev. D*, **6**, 798 (1972); H. Rechenberg and E. C. G. Sudarshan: *Nuovo Cimento*, **14** A, 299 (1973).

and the analytic properties of the scattering amplitudes of their bound states one cannot rule out the possibility of principal-value propagators for color gluons. In the case of quantum gravity, another potential application of principal-value propagators, the analytic structure of scattering amplitudes due to graviton exchange is unknown and likely to remain so. In the absence of such information, the renormalizability of pure quantum gravity, and the compatibility with Mach's principle, resulting from the use of principal-value propagators for gravitons is encouraging.

Perhaps the most important question facing field theory models with principal-value propagators is unitarity. The appendix contains a detailed discussion of this issue. The physical S-matrix is shown to be unitary. In addition the value of using infinite-momentum frame variables to compute S-matrix elements is pointed out.

2. – Model electrodynamics.

In this section we shall describe a model electrodynamics which is identical to quantum electrodynamics in its predictions. We shall also discuss a model for quantum action-at-a-distance electrodynamics.

The Lagrangian density of our models [1] is

$$(2)\qquad \mathscr{L} = -\tfrac{1}{2} F^1_{\mu\nu} F^{2\mu\nu} - \tfrac{1}{4} F^1_{\mu\nu} F^{1\mu\nu} + \tilde{\psi}_2 \gamma^0 (i(\gamma\cdot\nabla) - e_0(\gamma\cdot A_2) - m)\psi_2 - (\tilde{\psi}_1 - \tilde{\psi}_2)\gamma^0 (i(\gamma\cdot\nabla) - m)(\psi_1 - \psi_2)\,,$$

where we have introduced two electromagnetic fields, $A_{1\mu}$ and $A_{2\mu}$, so that

$$(3)\qquad F^i_{\mu\nu} = \partial_\nu A_{i\mu} - \partial_\mu A_{i\nu}\,,$$

for $i = 1, 2$, and two fields, ψ_1 and ψ_2 for electrons. The field $\tilde{\psi}_i$ is related to the transpose of the charge conjugate of ψ_i by [1]

$$(4)\qquad \tilde{\psi}_i = \psi_i^{cT} \gamma^0 C^T\,,$$

where C is a charge conjugation matrix. We follow the notation and conventions of ref. [5]. The free field theory for fermions has been developed in detail in ref. [1].

---

(5) J. BJORKEN and S. DRELL: *Relativistic Quantum Fields* (New York, N. Y., 1965).

Before discussing the free field theory of the photons we note the invariance of $\mathscr{L}$ under a restricted gauge transformation of the second kind,

$$\psi_2 \to \psi_2 \exp[i\Lambda], \tag{5}$$

$$\tilde{\psi}_2 \to \tilde{\psi}_2 \exp[-i\Lambda], \tag{6}$$

$$e_0 A_{2\mu} \to e_0 A_{2\mu} - \partial_\mu \Lambda, \tag{7}$$

$$\psi_1 \to \psi_1 + [\exp[i\Lambda] - 1]\psi_2, \tag{8}$$

$$\tilde{\psi}_1 \to \tilde{\psi}_1 + [\exp[-i\Lambda] - 1]\tilde{\psi}_2. \tag{9}$$

The associated conserved current is

$$J_\mu = \tilde{\psi}_2 \gamma^0 \gamma_\mu \psi_2. \tag{10}$$

It should be noted that the Lagrangian is also invariant under an independent gauge transformation, $A_{1\mu} \to A_{1\mu} - \partial_\mu \Lambda_1$.

Upon varying the action associated with $\mathscr{L}$, we find the field equations

$$\partial^\mu F^1_{\mu\nu} = \partial^\mu F^2_{\mu\nu}, \tag{11}$$

$$\partial^\nu F^2_{\mu\nu} = e_0 \tilde{\psi}_2 \gamma^0 \gamma_\nu \psi_2, \tag{12}$$

$$(i(\gamma\cdot\nabla) - m)\psi_1 = (i(\gamma\cdot\nabla) - m)\psi_2, \tag{13}$$

$$(i(\gamma\cdot\nabla) - e_0(\gamma\cdot A_2) - m)\psi_2 = 0. \tag{14}$$

Equations (12) and (14) demonstrate the equivalence of our model to the usual model of electrodynamics at the level of c-number fields.

We now turn to the case of free photons. From the Lagrangian we can identify the canonical momentum conjugate to $A_{1\mu}$ as

$$\pi_{1\mu} = F^2_{0\mu} - F^1_{0\mu}, \tag{15}$$

while the momentum conjugate to $A_{2\mu}$ is

$$\pi_{2\mu} = F^1_{0\mu}. \tag{16}$$

Since the free electromagnetic Lagrangian is invariant under independent gauge transformations of $A_{1\mu}$ and $A_{2\mu}$ the choice of gauges for $A_{1\mu}$ and $A_{2\mu}$ are also independent. We shall second quantize the fields in the joint Coulomb gauge

$$\vec{\nabla}\cdot\vec{A}_1 = \vec{\nabla}\cdot\vec{A}_2 = 0. \tag{17}$$

The resulting equal-time commutation relations are

$$[F^1_{0i}(\vec{x}, t), A_{1j}(\vec{y}, t)] = 0\,, \tag{18}$$

$$[F^1_{0i}(\vec{x}, t), A_{2j}(\vec{y}, t)] = i\delta^{tr}_{ij}(\vec{x} - \vec{y})\,, \tag{19}$$

$$[F^2_{0i}(\vec{x}, t), A_{2j}(\vec{y}, t)] = i\delta^{tr}_{ij}(\vec{x} - \vec{y})\,, \tag{20}$$

$$[F^2_{0i}(\vec{x}, t), A_{1j}(\vec{y}, t)] = i\delta^{tr}_{ij}(\vec{x} - \vec{y})\,, \tag{21}$$

with

$$\delta^{tr}_{ij}(\vec{x} - \vec{y}) = \int \frac{\mathrm{d}^3k}{(2\pi)^3} \exp[-i\vec{k}\cdot(\vec{x} - \vec{y})](\delta_{ij} - k_i k_j/|\vec{k}|^2)\,. \tag{22}$$

The field operator expansions are completely analogous to the expansion of a free scalar field discussed in ref. (1). The major difference is the necessary presence of polarization vectors in the electromagnetic-field expansions

$$\vec{A}_1(x) = \int \mathrm{d}^3k \sum_{\lambda=1}^{2} \vec{\varepsilon}(k, \lambda)[f_k(x)(\cos(\theta_1 - \theta_2)a_{1k\lambda} + \sin(\theta_1 - \theta_2)a_{2k\lambda})/\sin\theta_1 + f^*_k(x)(\cos(\theta_1 - \theta_2)a^\dagger_{1k\lambda} - \sin(\theta_1 - \theta_2)a^\dagger_{2k\lambda})/\cos\theta_1] \tag{23}$$

and

$$\vec{A}_2(x) = \int \mathrm{d}^3k \sum_{\lambda=1}^{2} \vec{\varepsilon}(k, \lambda)[f_k(x)(\cos\theta_2 a_{2k\lambda} + \sin\theta_2 a_{1k\lambda}) + f^*_k(x)(\sin\theta_2 a^\dagger_{2k\lambda} + \cos\theta_2 a^\dagger_{1k\lambda})]\,, \tag{24}$$

where θ_1 and θ_2 are arbitrary angles. As discussed in ref. (1) the choice of these angles represents a choice of boundary conditions for the free-field Green's functions. If the Fourier coefficient operators satisfy

$$[a_{ik\lambda}, a_{jk'\lambda'}] = [a^\dagger_{ik\lambda}, a^\dagger_{jk'\lambda'}] = 0\,, \tag{25}$$

$$[a_{ik\lambda}, a^\dagger_{jk'\lambda'}] = (1 - \delta_{ij})\delta_{\lambda\lambda'}\delta^3(\vec{k} - \vec{k}')\,, \tag{26}$$

for $i, j = 1, 2$, then the equal-time commutation relations will be satisfied for arbitrary θ_1 and θ_2. In addition the Hamiltonian which is the 0-0 component of the energy-momentum tensor derived from the Lagrangian, has the form

$$H = \int \mathrm{d}^3k \sum_{\lambda=1}^{2} \omega_k (a^\dagger_{2k\lambda} a_{1k\lambda} + a_{2k\lambda} a^\dagger_{1k\lambda})\,, \tag{27}$$

independent of θ_1 and θ_2. There is an analogy (6) between the quantization

(6) S. BLAHA: *Phys. Rev. D*, **17**, 994 (1978).

of the mode amplitudes, eqs. (25) and (26), and the co-ordinates of an assembly of one-dimensional harmonic oscillators just as in the second quantization of conventional field theory. We shall define photon vacua in the present case in a manner consistent with the procedure of ref. (1) and in analogy with the simple harmonic-oscillator model of ref. (6):

$$a_{1k\lambda}|0\rangle_2 = a^\dagger_{1k\lambda}|0\rangle_2 = 0\,, \tag{28}$$

$$a_{2k\lambda}|0\rangle_2 \neq 0\,, \qquad a^\dagger_{2k\lambda}|0\rangle_2 \neq 0\,, \tag{29}$$

for one vacuum, $|0\rangle_2$, and

$$a_{2k\lambda}|0\rangle_1 = a^\dagger_{2k\lambda}|0\rangle_1 = 0\,, \tag{30}$$

$$a_{1k\lambda}|0\rangle_1 \neq 0\,, \qquad a^\dagger_{1k\lambda}|0\rangle_1 \neq 0\,, \tag{31}$$

for the other vacuum, $|0\rangle_1$, for all k and λ. The vacua are related to each other by a Dirac-metric operator in a manner familiar from the scalar case. The metric operator γ is necessary in view of the commutation relations of the Fourier-coefficient operators, eqs. (25) and (26). A one-photon ket is defined by

$$|k\lambda\rangle = a^\dagger_{2k\lambda}|0\rangle_2\,, \tag{32}$$

while the dual one-photon bra is defined by

$$\langle k\lambda| = {}_1\langle 0|a_{1k\lambda}\,, \tag{33}$$

so that

$$\langle k'\lambda'|k\lambda\rangle = \delta_{\lambda\lambda'}\,\delta^3(\vec{k}-\vec{k}')\,, \tag{34}$$

by means of eq. (26) and ${}_1\langle 0|0\rangle_2 = 1$. The metric operator is introduced in order to relate the dual of a ket to its Hermitian conjugate:

$$\langle k\lambda| = (|k\lambda\rangle)^\dagger\gamma\,. \tag{35}$$

Consequently the metric operator is defined to have the properties

$$\gamma|0\rangle_1 = |0\rangle_2\,, \tag{36}$$

$$\gamma^{-1}a_{1k\lambda}\gamma = a_{2k\lambda}\,, \tag{37}$$

$$\gamma^{-1}a_{2k\lambda}\gamma = a_{1k\lambda} \tag{38}$$

with $\gamma = \gamma^\dagger = \gamma^{-1}$. It has a form similar to the metric operator of the scalar case which is given in eq. (36) of ref. (1). The definition of many-photon bras and kets is a direct generalization of eqs. (32), (33) and (35).

The time-ordered propagator which will later be of use in the development of perturbation theory is

$$(39)\qquad G^{22}_{\mu\nu}(x-y) = -i\,{}_1\langle 0|T(A_{2\mu}(x)A_{2\nu}(y))|0\rangle_2 = \\ = \sin^2\theta_2\, G_{F\mu\nu}(x-y) + \cos^2\theta_2\, G^*_{F\mu\nu}(x-y)\,,$$

where $G_{F\mu\nu}(x-y)$ is the usual Feynman propagator for photons, while G^*_{F} is its complex conjugate. Another nonzero free-field propagator, $G^{12}_{\mu\nu}$ is defined as the vacuum expectation value of the time-ordered product of $A_{1\mu}(x)$ and $A_{2\nu}(y)$. It depends on both θ_1 and θ_2. It does not appear in perturbation theory.

Equation (39) shows that $G^{22}_{\mu\nu}$ is a Feynman propagator if $\theta_2 = \pi/2$, and a principal-value propagator if $\theta_2 = \pi/4$. Thus

$$(40)\qquad G^{22}_{\mu\nu}(x-y)|_{\theta_2=\pi/4} = g_{\mu\nu}\int \frac{d^4k}{(2\pi)^4}\exp[-ik\cdot(x-y)]\,\frac{P}{k^2}\,,$$

where

$$(41)\qquad \frac{P}{k^2} = \frac{1}{2}\left(\frac{1}{k^2+i\varepsilon}+\frac{1}{k^2-i\varepsilon}\right),$$

in the Feynman gauge. This Green's function has previously appeared in action-at-a-distance formulations (3) of classical electrodynamics. The generality of our formulation of quantum field theory allows it to appear as a special case. This does not seem to be possible within the framework of the conventional formulation (7). If we choose $\theta_2 = \pi/4$ and proceed to develop perturbation theory for the full interacting theory then we obtain a quantum-action-at-a-distance theory. We obtain quantum exchange of energy and momentum to which we associate lines in Feynman diagrams. The only difference is in the analytic structure associated with the « epsilontics » of the pole locations. Scattering amplitudes are piecewise analytic in this case (4). More importantly, quanta associated with the electromagnetic field do not appear in asymptotic states, if unitarity is to be maintained. Thus the electromagnetic field does not have its own degrees of freedom, but is constrained to have its source in the charged matter fields which are present. The physical expectations of a quantum action-at-a-distance electrodynamics are therefore satisfied. Except for the absence of « photons » from asymptotic states, and the principal-value nature of the electromagnetic-field propagator, the perturbation theory rules, diagrams and combinatorics are identical to those of QED, although they must be unitarized using the method described in the appendix.

---

(7) R. FEYNMAN: *Science*, **153**, 699 (1966).

We now turn to the Feynman case ($\theta_2 = \pi/2$). The field $\vec{A}_2$ takes the form

$$\vec{A}_2 = \int d^3k \sum_{\lambda=1}^{2} \vec{\varepsilon}(k, \lambda)[f_k(x) a_{1k\lambda} + f_k^*(x) a_{2k\lambda}^\dagger], \tag{42}$$

in the Coulomb gauge. Note that the form of eq. (42) and the commutators, eq. (25) and (26) imply the free-field identity

$$_1\langle 0|A_{2i_1}(x_1) A_{2i_2}(x_2) \dots A_{2i_n}(x_n)|0\rangle_2 = \langle 0|A_{i_1}(x_1) A_{i_2}(x_2) \dots A_{i_n}(x_n)|0\rangle \tag{43}$$

for all spatial indices $i_1, i_2, \dots, i_n$, and all n for which the right-hand side is the vacuum expectation value of n free fields in the usual formulation. This identity is the basis of our claim that the Feynman case of our formulation is *completely* equivalent to QED in its predictions. An important part of establishing the correspondence is the LSZ reduction formulae for photons: for an in photon $(k\lambda)$ we have

$$\begin{aligned} \langle \beta \text{ out}|\alpha(k\lambda) \text{ in}\rangle &= \langle \beta \text{ out}|a_{2k\lambda \text{in}}^\dagger|\alpha \text{ in}\rangle = \\ &\langle \beta - (k\lambda) \text{ out}|\alpha \text{ in}\rangle - \frac{i}{\sqrt{Z_3}} \int d^4x\, A_{k\lambda}^{\mu}(x) \overrightarrow{\Box} \langle \beta \text{ out}|A_{2\mu}(x)|\alpha \text{ in}\rangle \end{aligned} \tag{44}$$

and for an out photon $(k\lambda)$ we have

$$\begin{aligned} \langle \beta(k\lambda) \text{ out}|\alpha \text{ in}\rangle &= \langle \beta \text{ out}|a_{1k\lambda \text{ out}}|\alpha \text{ in}\rangle = \\ &= \langle \beta \text{ out}|\alpha - (k\lambda) \text{ in}\rangle - \frac{i}{\sqrt{Z_3}} \int d^4x \langle \beta \text{ out}|A_{2\mu}(x)|\alpha \text{ in}\rangle \overleftarrow{\Box} A_{k\lambda}^{\mu *}(x) \end{aligned} \tag{45}$$

in the notation of ref. ([5]).

The second quantization of the free-electron field is described in detail in ref. ([1]). We assume an in and out electron field formalism is developed along those lines. Our formulation of spin-one-half fermion theory also allows one to choose Feynman propagators or principal-value propagators for fermions. In order to establish a model equivalent to QED we choose Feynman propagators for the electrons.

The perturbation theory for our model is based on the interaction Lagrangian

$$\mathscr{L}_{\text{int}} = - e_0 \tilde{\psi}_2 \gamma^0 (\gamma \cdot A_2) \psi_2 . \tag{46}$$

We shall need the LSZ reduction formulae for electrons

$$\begin{aligned} \langle \beta \text{ out}|(ps)\alpha \text{ in}\rangle &= \langle \beta - (ps) \text{ out}|\alpha \text{ in}\rangle - \\ &- \frac{i}{\sqrt{Z_2}} \int d^4x \langle \beta \text{ out}|\tilde{\psi}_2(x)\gamma^0|\alpha \text{ in}\rangle (-i\overleftarrow{\nabla} - m)\, U_{ps}(x) \end{aligned} \tag{47}$$

and

$$(48)\qquad \langle\beta(ps)\,\text{out}|\alpha\,\text{in}\rangle = \langle\beta\,\text{out}|\alpha\quad(ps)\,\text{in}\rangle - -\frac{i}{\sqrt{Z_2}}\int d^4x\,\overline{U}_{ps}(x)(i\overrightarrow{\nabla}-m)\langle\beta\,\text{out}|\psi_2(x)|\alpha\,\text{in}\rangle$$

in the notation of ref. (5). To remove a positron from an in state, we use

$$(49)\qquad \langle\beta\,\text{out}|(ps)\alpha\,\text{in}\rangle = \langle\beta-(ps)\,\text{out}|\alpha\,\text{in}\rangle + +\frac{i}{\sqrt{Z_2}}\int d^4x\,\overline{V}_{\overline{ps}}(x)(i\overrightarrow{\nabla}-m)\langle\beta\,\text{out}|\psi_2(x)|\alpha\,\text{in}\rangle$$

and, to remove a positron from an out state, we use

$$(50)\qquad \langle\beta(ps)\,\text{out}|\alpha\,\text{in}\rangle = \langle\beta\,\text{out}|\alpha-(ps)\,\text{in}\rangle + +\frac{i}{\sqrt{Z_2}}\int d^4x\langle\beta\,\text{out}|\overline{\psi}_2\gamma^0|\alpha\,\text{in}\rangle(-i\overleftarrow{\nabla}-m)\,V_{\overline{ps}}(x)\,.$$

In view of the interaction Lagrangian and eqs. (47)-(50) it is clear that only time-ordered products of $\psi_{2\text{in}}$ and $\overline{\psi}_{2\text{in}}$ appear in the perturbation theory expansion. The Wick theorem, which applies in the present case, reduces all vacuum expectation values of time-ordered products of $\psi_{2\text{in}}$ and $\overline{\psi}_{2\text{in}}$ to products of the time-ordered two-point function

$$(51)\qquad iS(x-y) = {}_1\langle 0|T(\psi_2(x)\,\overline{\psi}_2(y)\gamma^0)\,|0\rangle_2$$

$$(52)\qquad = i\int\frac{d^4p}{(2\pi)^4}\exp[-ip\cdot(x-y)]\,\frac{p+m}{p^2-m^2+i\varepsilon}\,,$$

where eqs. (51) is in the notation of ref. (1).

The above considerations and the familiar development* of the S-matrix as an expansion in vacuum expectation values of time-ordered products of in-field operators lead us to assert that our model electrodynamics is identical to QED in its predictions. In particular

$$(53)\qquad {}_1\langle 0|T\Big(A_{2\mu_1\text{in}}(x_1)A_{2\mu_2\text{in}}(x_2)\ldots\psi_{2\text{in}}(y_1)\psi_{2\text{in}}(y_2)\ldots\overline{\psi}_{2\text{in}}(z_1)\overline{\psi}_{2\text{in}}(z_2)\ldots \ldots\exp\Big[i\int_{-\infty}^{\infty}dt\,L_{\text{int}}\Big]\Big)|0\rangle_2 =$$

$$= \langle 0|T\Big(A_{\mu_1\text{in}}(x_1)A_{\mu_2\text{in}}(x_2)\ldots\psi_{\text{in}}(y_1)\psi_{\text{in}}(y_2)\ldots\overline{\psi}_{\text{in}}(z_1)\overline{\psi}_{\text{in}}(z_2)\ldots\exp\Big[i\int_{-\infty}^{\infty}dt\,L_{\text{int}}^{\text{QED}}\Big]\Big)|0\rangle\,,$$

where the right-hand side is the corresponding time-ordered product of the conventional formulation of QED with

$$L^{\text{QED}}_{\text{int}} = -i : \int \mathrm{d}^3x\, e_0 \bar{\psi}_{\text{in}}(\gamma\cdot A_{\text{in}})\psi_{\text{in}} : . \tag{54}$$

Thus, the perturbation theory of our model electrodynamics is term by term equivalent to QED—the equivalent QED term being obtained by ignoring the subscripts 1 and 2 on fields and vacua, and letting $\tilde{\psi}_2\gamma^0 \to \bar{\psi}$. Feynman diagrams can be appropriated without modification for use in our formulation. We conclude with a quote from Feynman ([7]): « It always seems odd to me that the fundamental laws of physics can appear in so many different forms that are not apparently identical at first.... Theories of the known which are described by different physical ideas may be equivalent in all their predictions and hence scientifically indistinguishable. However they are not psychologically identical when one is trying to move from that base into the unknown. For different views suggest different modifications ... » In that spirit we turn to the investigation of non-Abelian theories.

3. – Yang-Mills models.

The models we have hitherto investigated are members of a class whose Lagrangians have the form

$$\mathscr{L} = \mathscr{L}_0(\varphi_2) + \mathscr{L}_1(\varphi_1 - \varphi_2) , \tag{55}$$

where $\mathscr{L}_0$ has a form which is essentially identical to a Lagrangian of the conventional formulation. For example, in the case of electrodynamics the Lagrangian of eq. (2) can be put in the form of eq. (55), if we let

$$\mathscr{L}_0(A_{2\mu}, \psi_2) = -\tfrac{1}{4} F^2_{\mu\nu} F^{2\mu\nu} + \tilde{\psi}_2\gamma^0\big(i(\gamma\cdot\nabla) - e_0(\gamma\cdot A_2) - m\big)\psi_2 \tag{56}$$

and

$$\mathscr{L}_1(A_{1\mu} - A_{2\mu}, \psi_1 - \psi_2) = \\ = -\tfrac{1}{4}(F^1_{\mu\nu} - F^2_{\mu\nu})^2 + (\tilde{\psi}_2 - \tilde{\psi}_1)\gamma^0\big(i(\gamma\cdot\nabla) - m\big)(\psi_1 - \psi_2) . \tag{57}$$

At the level of c-number fields it is clear from eqs. (56) and (57) that $A_{2\mu}$ and ψ_2 reproduce the usual electrodynamics. At the level of quantum fields one can opt to have a model reproducing the conventional quantum theory—we call it the Feynman case. It maintains the implied decoupling of $A_{1\mu} - A_{2\mu}$ and $\psi_1 - \psi_2$ from $A_{2\mu}$ and ψ_2. On the other hand, we can take advantage of the larger space of states in the larger-manifestly-indefinite-metric theory and couple the fields together by taking advantage of degrees of freedom in the

Fourier expansion of asymptotic fields and the definition of the vacuum. Since we « perturb around » the asymptotic fields, perturbation theory will reflect this coupling. Our principal-value or quantum action-at-a-distance models are based on this possibility.

We shall use the form of eq. (55) as an ansatz to generate Yang-Mills models in our formulation. This leads to the gauge field Lagrangian

$$\mathscr{L} = -\tfrac{1}{4}\boldsymbol{F}_{2\mu\nu}\cdot\boldsymbol{F}_2^{\mu\nu} + \tfrac{1}{4}(\boldsymbol{G}_{1\mu\nu} - \boldsymbol{G}_{2\mu\nu})\cdot(\boldsymbol{G}_1^{\mu\nu} - \boldsymbol{G}_2^{\mu\nu}), \tag{58}$$

where

$$\boldsymbol{F}_{2\mu\nu} = \partial_\mu \boldsymbol{A}_{2\nu} - \partial_\nu \boldsymbol{A}_{2\mu} + g\boldsymbol{A}_{2\mu}\times\boldsymbol{A}_{2\nu} \tag{59}$$

and

$$\boldsymbol{G}_{i\mu\nu} = \partial_\mu \boldsymbol{A}_{i\nu} - \partial_\nu \boldsymbol{A}_{i\mu} \tag{60}$$

for $i = 1, 2$. Under a local gauge transformation, $S = S(x)$,

$$A_{2\mu} \to A'_{2\mu} = S^{-1}A_{2\mu}S + \frac{i}{g}S^{-1}\partial_\mu S, \tag{61}$$

$$A_{1\mu} \to A'_{1\mu} = A_{1\mu} + A'_{2\mu} - A_{2\mu}, \tag{62}$$

the Lagrangian $\mathscr{L}$ is invariant, where $A_{i\mu} = \boldsymbol{A}_{i\mu}\cdot\boldsymbol{T}$ for $i = 1, 2$ with $\boldsymbol{T}$ a vector composed of matrices representing generators of the gauge group. There is an additional invariance under the local transformation, $\boldsymbol{A}_{1\mu} \to \boldsymbol{A}_{1\mu} + \partial_\mu\boldsymbol{\Lambda}$, which is a straightforward generalization of the Abelian gauge transformation.

The field equations derived from eq. (58), upon varying the action with respect to $\boldsymbol{A}_{1\mu}$ and $\boldsymbol{A}_{2\mu}$, are

$$\partial^\mu(\boldsymbol{G}_{1\mu\nu} - \boldsymbol{G}_{2\mu\nu}) = 0 \tag{63}$$

and

$$(\partial^\mu + g\boldsymbol{A}_2^\mu\times)\boldsymbol{F}_{2\mu\nu} = 0\,. \tag{64}$$

The canonical momentum conjugate to $\boldsymbol{A}_{2\mu}$ is

$$\boldsymbol{\Pi}_{2\mu} = \boldsymbol{F}_{2\mu0} + \boldsymbol{G}_{1\mu0} - \boldsymbol{G}_{2\mu0} \tag{65}$$

and the momentum conjugate to $\boldsymbol{A}_{1\mu}$ is

$$\boldsymbol{\Pi}_{1\mu} = \boldsymbol{G}_{2\mu0} - \boldsymbol{G}_{1\mu0}\,. \tag{66}$$

We choose the Coulomb gauge to implement the field quantization. Due to the form of the Lagrangian, we can treat $\boldsymbol{A}_{2\mu}$ and $\boldsymbol{A}_{3\mu} = \boldsymbol{A}_{1\mu} - \boldsymbol{A}_{2\mu}$ as independent fields. Choosing the Coulomb gauge for $A_{2\mu}$, $\vec{\nabla}\cdot\vec{A}_2 = 0$, we can isolate the independent field quantities and establish their equal-time commutation relations in the well-known manner [8]. We also can choose to work in the Coulomb gauge of $A_{3\mu}$, $\vec{\nabla}\cdot\vec{A}_3 = 0$ which is equivalent to $\vec{\nabla}\cdot\vec{A}_1 = 0$. This possibility follows from the invariance of $\mathscr{L}$ under the local « Abelian » transformation of $A_{3\mu}$ or $A_{1\mu}$ mentioned above. The resulting equal-time commutation relations are equivalent to

$$[G^{\mathrm{T}}_{\alpha 0 i a}(x), A_{\beta j b}(y)] = i(1 - \delta_{\alpha 1}\delta_{\beta 1})\delta_{ab}\Delta^{\mathrm{tr}}_{ij}(x - y)\,, \tag{67}$$

where $\alpha, \beta = 1, 2$; i and j are spatial- and a and b are internal-symmetry indices. G^{T} is the transverse part of G. One can now proceed to linearize the field equations. The Fourier expansions of the solutions of the linearized equations have the form of eqs. (23) and (24), if an internal-symmetry index is appended to the Fourier component operators. The discussion then reduces to the Abelian case considered in the last section. In particular the Green's function of the linearized model which is relevant to perturbation theory is

$$G^{22}_{ab\mu\nu}(x - y) = -i\,{}_1\langle 0|T\big(A_{2\mu a}(x)\,A_{2\nu b}(y)\big)|0\rangle_2\,. \tag{68}$$

It can be expressed as a linear combination of a Feynman propagator and its complex conjugate as in eq. (39). The arbitrary angle can be chosen to give Feynman propagators or principal-value propagators.

The perturbation theory is based on the linearized model. In the Feynman case the LSZ reduction of asymptotic non-Abelian quanta is essentially given in eqs. (44) and (45), if appropriate internal-symmetry indices are introduced. Only fields of « type 2 » are introduced by the LSZ reduction. Furthermore, the cubic and quartic terms of the Lagrangian « interaction » term also involve only fields of type 2. If we restrict all couplings to other fields—such as fermions—to involve only $A_{2\mu}$, then only time-ordered products of $A_{2\mu}$ appear in the perturbation theory expansion of the S-matrix. The consequence of this development is that one has constructed a model which makes predictions which are identical to a conventional formulation based on a Lagrangian of the usual type

$$\mathscr{L} = -\tfrac{1}{4}\boldsymbol{F}_{\mu\nu}\cdot\boldsymbol{F}^{\mu\nu}\,. \tag{69}$$

This can be verified in a path integral approach based on the Lagrangian of eq. (58).

The point of our formulation is that one can make other choices—such as the principal-value propagator choice. (Note that the path integral formulation

---

(8) E. ABERS and B. W. LEE: *Phys. Rep.*, **9** C, 1 (1973).

only has a formal validity in these cases, since the Wick rotation to Euclidean co-ordinates is highly nontrivial.) The choice of principal-value propagators has several important consequences.

In the appendix we show how to calculate the unitary physical S-matrix in this case. We show that loops of principal-value propagators do not contribute to the absorptive parts of S-matrix elements. As a result Faddeev-Popov ghost loops are not needed to maintain unitarity (in any gauge). We also point out a radical possibility: a unitary S-matrix can be defined in which all non-Abelian boson loops are excluded. Thus the non-Abelian boson sector consists solely of tree diagrams. This is a quantum analogue of the absence of self-interactions in classical action-at-a-distance theories.

4. – Non-Abelain model of the strong interactions.

Several years ago a non-Abelian model of the strong interaction was constructed (9) which had manifest quark confinement and a linear potential. The model was based on higher-order field equations. Such equations lead to well-known unitarity problems, if the usual quantization program is implemented. The author chose to avoid the unitarity problems through an *ad hoc* quantization procedure which gave principal-value propagators. The result was a model having unconventional analyticity properties. Despite the successful confinement scheme in this model, there was some reason for uneasiness, due to the unusual form of the Lagrangian (where two sets of fields were associated with the gluons) and the *ad hoc* method of quantization.

The new framework we have developed in ref. (1) and this paper eliminates these sources of concern. Our formulation is based on fundamental ground—the need for an acceptable physical particle interpretation of quantum field theory in flat and curved space-time. We shall see that the two-field Lagrangian of the strong-interaction model lies within the general framework we have established. The principal-value propagators, required to have a unitary model, will also be obtained in a straightforward manner.

Since the strong-interaction model is described in some detail in ref. (9), we shall only outline the aspects necessary to connect that model with the present work.

The strong-interaction Lagrangian departs slightly from the ansatz of eq. (55) (mainly in the replacement $\boldsymbol{G}^1_{\mu\nu}\cdot\boldsymbol{G}^{1\mu\nu}\to\Lambda^2\boldsymbol{A}_{1\mu}\cdot\boldsymbol{A}^\mu_1$):

$$(70)\qquad \mathscr{L}=-\tfrac{1}{2}\boldsymbol{F}^1_{\mu\nu}\cdot\boldsymbol{F}^{2\mu\nu}-\tfrac{1}{2}\Lambda^2\boldsymbol{A}_{1\mu}\cdot\boldsymbol{A}^\mu_1+\tilde{\psi}_2\gamma^0\big(i(\gamma\cdot\nabla)+g(\gamma\cdot\boldsymbol{A}_2)-m\big)\psi_2-(\tilde{\psi}_1-\tilde{\psi}_2)\gamma^0\big(i(\gamma\cdot\nabla)-m\big)(\psi_1-\psi_2)\,.$$

(9) S. BLAHA: *Phys. Rev. D*, **10**, 4268 (1974); **11**, 2921 (1975). It should be noted that the claim that loops of principal-value propagators are identically zero is not true. In the case of one-loop diagrams only the real part is zero.

where ψ_1 and ψ_2 are the quark fields and

$$F^2_{\mu\nu} = \partial_\mu \boldsymbol{A}_{2\nu} - \partial_\nu \boldsymbol{A}_{2\mu} + g\boldsymbol{A}_{2\mu} \times \boldsymbol{A}_{2\nu} \tag{71}$$

and

$$F^1_{\mu\nu} = D_\mu \boldsymbol{A}_{1\nu} - D_\nu \boldsymbol{A}_{1\mu} \tag{72}$$

with the covariant derivative defined by

$$D_\mu = \partial_\mu + g\boldsymbol{A}_{2\mu} \times . \tag{73}$$

The Lagrangian is invariant under the local color SU_3 gauge transformation

$$\psi_2 \to S^{-1}\psi_2 , \tag{74}$$

$$\psi_1 \to \psi_1 + (S^{-1} - I)\psi_2 , \tag{75}$$

$$A_{1\mu} \to S^{-1} A_{1\mu} S , \tag{76}$$

$$A_{2\mu} \to S^{-1} A_{2\mu} S + \frac{i}{g} S^{-1} \partial_\mu S , \tag{77}$$

with $A_{i\mu} = \boldsymbol{A}_{i\mu} \cdot \boldsymbol{T}$, where $\boldsymbol{T}$ is a vector composed of matrices representing SU_3 generators. The equations of motion and the Coulomb gauge quantization are discussed in detail in ref. [9]. (The subscripts, 1 and 2, on the gluon fields are reversed in ref. [9] relative to the present development.)

The essential feature of the model can be seen in the linearized field equations

$$\Box \boldsymbol{A}_{2\mu} = -\Lambda^2 \boldsymbol{A}_{1\mu} , \tag{78}$$

$$\Box \boldsymbol{A}_{1\mu} = - g\boldsymbol{J}_\mu , \tag{79}$$

where $\boldsymbol{J}_\nu$ is the color quark current. Equations (78) and (79) show an apparent mass term, $\Lambda^2(\boldsymbol{A}_{1\mu})^2$, actually leads to a higher-order derivative field equation. The implications of this feature for quark confinement are discussed in ref. [9]. The discussion is based on the assumption that all gluon propagators are principal-value propagators. We shall now show that the propagators for color gluons in the linearized model can be principal-value propagators.

We work in the Coulomb gauge, $\vec{\nabla} \cdot \vec{\boldsymbol{A}}_2 = 0$. A mode expansion of the gluon fields must be consistent with the free linearized field equations and the equal-time commutation relations,

$$[F^{\lambda T}_{0ia}(x), A_{\beta jb}(y)] = i\delta_{ab}(1 - \delta_{\lambda\beta})\delta^{tr}_{ij}(\vec{x} - \vec{y}) \tag{80}$$

(cf. eqs. (65) and (66) of ref. [9]). These conditions and the requirement of a unitary theory imply the Fourier expansions

$$\vec{A}_{1a}(x) = \int \frac{\mathrm{d}^3k}{\sqrt{2}} \sum_{\lambda=1}^{2} \vec{\varepsilon}(k,\lambda)[(a_{1k\lambda a} + a_{2k\lambda a})f_k(x) + (a^\dagger_{1k\lambda a} + a^\dagger_{2k\lambda a})f^*_k(x)], \tag{81}$$

$$\vec{A}_{2a}(x) = \int \frac{\mathrm{d}^3k}{\sqrt{2}} \sum_{\lambda=1}^{2} \vec{\varepsilon}(k,\lambda)[(a_{2k\lambda a} - a_{1k\lambda a})f_k(x) + (a^\dagger_{2k\lambda a} - a^\dagger_{1k\lambda a})f^*_k(x)] + \tag{82}$$

$$+ \Lambda^2 \int \frac{\mathrm{d}^4k}{\sqrt{2}} 2\omega_k \theta(k_0)\delta'(k^2) \sum_{\lambda=1}^{2} \vec{\varepsilon}(k,\lambda)[(a_{1k\lambda a} + a_{2k\lambda a})f_k(x) + (a^\dagger_{1k\lambda a} - a^\dagger_{2k\lambda a})f^*_k(x)],$$

where $\delta'(k^2) = \partial\delta(k^2)/\partial k^2$ and $\omega_k = |\vec{k}|$. From eqs. (81) and (82) we can determine the Green's functions of the linearized model:

$$iG^{12}_{\mu\nu ab}(x-y) = {}_1\langle 0|T(A_{1\mu a}(x)A_{2\nu b}(y))|0\rangle_2 = \tag{83}$$

$$= -i\delta_{ab}\int \frac{\mathrm{d}^4k}{(2\pi)^4} \exp[-ik\cdot(x-y)] r^{12}_{\mu\nu} P\frac{1}{k^2}$$

and

$$iG^{22}_{\mu\nu ab}(x-y) = {}_1\langle 0|T(A_{2\mu a}(x)A_{2\nu b}(y))|0\rangle_2 = \tag{84}$$

$$= i\Lambda^2\delta_{ab}\int \frac{\mathrm{d}^4k}{(2\pi)^4} \exp[-ik\cdot(x-y)] r^{22}_{\mu\nu} P\frac{1}{k^4},$$

where

$$P\frac{1}{k^{2N}} = \frac{1}{2}\left[\frac{1}{(k^2+i\varepsilon)^N} + \frac{1}{(k^2-i\varepsilon)^N}\right] \tag{85}$$

and where $r^{12}_{\mu\nu}$ and $r^{22}_{\mu\nu}$ are gauge-dependent tensors. While the path integral formulation is only of formal value for the present model, it can be used to determine the form of the Green's functions for any choice of gauge. It can also be used to generate the perturbation theory rules for the model. The character of the perturbation theory is described in ref. (9). Confinement arises through the Schwinger mechanism in a manner reminiscent of the two-dimensional Schwinger model. The mass scale is set by Λ. The Lagrangian of eq. (70) leads to an interaction between quarks which has the linear potential r as its Coulomb potential in lowest order. The r potential has had notable success in the explication of the charmonium system. A r^{-1} term may also be needed to successfully describe charmonium. Such a term can be introduced in our model by adding another interaction term

$$\mathcal{L}_c = g'\bar{\psi}_2\gamma^0(\gamma\cdot A_1)\psi_2 \tag{86}$$

to eq. (70). Since $A_{1\mu}$ transforms homogeneously under a gauge transformation, the gauge invariance of the Lagrangian is not altered by the addition of this term.

In conclusion, we have established the basis for a manifestly quark-confining model of the strong interaction within the framework of our formulation of quantum field theory. This example illustrates the value of our formulation in the construction of unitary quantum field theories with higher-order derivative field equations. Another interesting application of this approach would be to quantum gravity which may have its renormalization problems ameliorated by introducing higher-order derivatives. Certain formal similarities, and this possibility, have led the author to propose a unified model of vierbein gravity and the strong interaction (10). The higher-order derivative gravity terms which lead to power counting renormalizability for the quantum gravity sector are linked to the higher-order derivative strong-interaction sector terms which lead to quark confinement. The recent discovery of important spin effects in high-energy p-p scattering (11) is suggestive in view of the presence of strong spin-spin interactions in unified models of the strong-interaction and vierbein gravity.

5. – Model of quantum gravity.

In this section we develop a trivially renormalizable model of quantum gravity (based on the Einstein Lagrangian) which is coupled to an external classical source. In addition to obtaining a physically interesting model, we shall see the utility of principal-value propagators in ameliorating divergence problems. The self-interactions of the « gravitons » will be reduced if principal-value propagators are used. This corresponds to the absence of self-interactions in classical action-at-a-distance models.

The model which we shall develop lies within the framework established above. It can be described as quantum action-at-a-distance gravity. We begin with the Einstein action

$$I = -2\varkappa^{-2}\int \mathrm{d}^4x\, g^{\frac{1}{2}} R \tag{87}$$

with $\varkappa = 32\pi G$, where G is Newton's constant, $g = \det g_{\mu\nu}$, $R = g^{\mu\nu}R^{\lambda}_{\mu\nu\lambda}$ and where

$$R^{\alpha}_{\mu\nu\lambda} = \partial_\nu \Gamma^{\alpha}_{\mu\lambda} + \Gamma^{\alpha}_{\nu\beta}\Gamma^{\beta}_{\mu\lambda} - (\nu \Leftrightarrow \lambda) \tag{88}$$

and

$$\Gamma^{\alpha}_{\mu\nu} = \frac{1}{2} g^{\alpha\beta}[\partial_\mu g_{\nu\beta} + \partial_\nu g_{\mu\beta} - \partial_\beta g_{\mu\nu}]\,. \tag{89}$$

---

(10) S. Blaha: *Lett. Nuovo Cimento*, **18**, 60 (1977).

(11) J. R. O'Fallon, L. G. Ratner, P. F. Schultz, K. Abe, R. C. Fernow, A. D. Krisch, T. A. Mulera, A. J. Satthouse, B. Sandler, K. M. Terwilliger, D. G. Crabb and P. H. Hansen: *Phys. Rev. Lett.*, **39**, 733 (1977).

Our approach is analogous to Feynman's covariant quantization around a flat background field (12). We therefore let

$$g_{\mu\nu} = \eta_{\mu\nu} + \varkappa h^2_{\mu\nu}, \tag{90}$$

where $\eta_{\mu\nu} = (-1, 1, 1, 1)$. We then introduce a second tensor field, $h^1_{\mu\nu}$; and choose the action for our model to be (cf. eq. (55))

$$I = \int d^4x [\tfrac{1}{2}(h^3_{\mu\nu,\sigma})^2 - (h^3_\mu)^2 - \tfrac{1}{2}(h^3_{,\mu})^2 - 2\varkappa^{-2} g^{\frac{1}{2}} R], \tag{91}$$

where $h^3_{\mu\nu} = h^1_{\mu\nu} - h^2_{\mu\nu}$, $h^3_\mu = h^3_{\mu\nu,\nu}$, $h^3_{,\mu} = h_{\nu\nu,\mu} = \partial_\mu h$ and where the last term on the right-hand side of eq. (91) is expanded in $h^2_{\mu\nu}$ by using eq. (90). The gauge invariance of the Lagrangian is maintained by requiring $h^1_{\mu\nu}$ and $h^2_{\mu\nu}$ to satisfy the same transformation law

$$h^i_{\mu\nu} \to h'^i_{\mu\nu} = h^i_{\mu\nu} - \partial_\nu \varepsilon_\mu - \partial_\mu \varepsilon_\nu, \tag{92}$$

where ε_μ are four small, but otherwise arbitrary, functions of the co-ordinates. We introduce a de Donder harmonic gauge fixing term

$$\mathscr{L}_D = -\frac{1}{2\gamma}\left(h^2_\mu - \frac{1}{2}h^2_{,\mu}\right)^2 + \frac{1}{2\gamma}\left(h^3_\mu - \frac{1}{2}h^3_{,\mu}\right)^2, \tag{93}$$

in order to obtain a regular kinetic matrix from the quadratic terms of the augmented Lagrangian. The quadratic terms are (if we let $\gamma = \frac{1}{2}$)

$$\mathscr{L}_{quad} = -h^1_{\alpha\beta,\mu} h^2_{\alpha\beta,\mu} + \tfrac{1}{2} h^1_{,\mu} h^2_{,\mu} + \tfrac{1}{2}(h^1_{\alpha\beta,\mu})^2 - \tfrac{1}{4}(h^1_{,\mu})^2. \tag{94}$$

These terms plus the higher-order «interaction» terms can be introduced into a suitable path integral (13) expression and the perturbation theory rules can be developed. The path integral approach is only of formal significance, in general, in our models. It gives the correct algebraic expressions and combinational rules. The nontriviality of Wick rotation to Euclidean co-ordinates in any case, but the Feynman case, precludes attributing anything more than formal value to the path integral approach. Of course, the applicability of the path integral formalism to the Feynman case, and the absence of any difference at the algebraic level (before integrations) between the Feynman and principal-value cases, means that the path integral formalism also describes the combinatorics of the principal-value case.

---

(12) R. Feynman: *Acta Phys. Polonica*, **24**, 697 (1963).
(13) V. N. Popov and L. Fadeev: Kiev Report. No. ITP 67-36 (1967).

In order to have a nontrivial model in the principal-value case, we shall assume the gravitational field is coupled to an external classical source. Note that covariance requires that only $h^2_{\mu\nu}$ couples to that source. This fact, plus the absence of $h^1_{\mu\nu}$ in higher-order terms, implies that the only free-field propagator necessary to develop perturbation theory is the vacuum expectation value of the time-ordered product of free $h^2_{\mu\nu}$ fields:

$$(95)\qquad iG^{22}_{\mu\nu,\varrho\sigma} = {}_1\langle 0|T\big(h^2_{\mu\nu}(x)h^2_{\varrho\sigma}(y)\big)|0\rangle_2 = \\ = -\frac{i}{2}(\eta_{\mu\varrho}\eta_{\nu\sigma} + \eta_{\mu\sigma}\eta_{\nu\varrho} - \eta_{\mu\nu}\eta_{\varrho\sigma})\int\frac{d^4k}{(2\pi)^4}\frac{\exp[-ik\cdot(x-y)]}{k^2}\,.$$

If one chooses the propagator in eq. (95) to be a Feynman propagator, then the usual model of quantum gravity results with its attendant renormalizability problems.

We shall consider the case, in which G^{22} is a principal-value propagator. One can establish this case by following a canonical quantization procedure which is completely analogous to those of models considered earlier. The propagation of « gravitons » by principal-value propagators has important consequences for perturbation theory. All graviton loops can be excluded from the physical S-matrix without leading to unitarity problems (cf. appendix). As a result, only tree diagrams occur and unitarity requires that all gravitons are absorbed—either within trees or on the external classical source. Thus we have a model quantum gravity with no renormalization problem. In the classical limit the model becomes Einstein's classical theory of gravity ([14]). There is, of course, the question of whether the classical limit is a retarded model of gravity or not. The answer lies in cosmology—is the absorber mechanism operative?

An action-at-a distance model of gravity with absorber seems to be very much in the spirit of Mach's principle. The usual approach—where the gravitational field is treated as having its own degrees of freedom—almost invariably leads to the surreptitious introduction of absolute space. The reason is simple. The solution of the field equations for a localized mass distribution is asymptotically flat. The only apparent way to avoid this problem is to say Mach restricts us to the class of closed-universe solutions.

In an action-at-a-distance gravity ([15]) the metric field exists only in the presence of matter, so that any asymptotic flatness of solutions is a mathematical

([14]) There is an opinion which is sometimes expressed that a tree diagram model is equivalent to a classical theory. This is not true. The correct view is that the classical limit of a tree diagram model is the corresponding classical theory. Cf. S. DESER's paper in *Quantum Gravity*, edited by C. J. ISHAM, R. PENROSE and D. W. SCIAMA (Oxford, 1975).

([15]) A. WHITEHEAD: *The Principle of Relativity* (Cambridge, 1922).

artifact. The metric properties of space are a consequence of the presence of mass-energy. (Consistent with this view, we note that mass-energy must be present to measure the metric properties of space.)

The ultimate realization of Mach's principle from the present viewpoint requires that inertia be the result of the gravitational effects of distant matter. Preliminary work ([16]) in this direction can be easily incorporated within the framework of an action-at-a-distance model of gravity. In fact, Sciama's model of the inertial effects of the distant stars displays a close analogy to the Feynman-Wheeler absorber model.

In closing we remark that the introduction of quantum matter fields in our action-at-a-distance quantum gravity appears to result in a nonrenormalizable model. It seems that a higher-order derivative Lagrangian of the type mentioned in the last section may be necessary for a fully renormalizable quantum gravity. Of course, that would also be a quantum action-at-a-distance model.

APPENDIX

In this appendix we define a unitary physical S-matrix for action-at-a-distance quantum field theories. The usual formal definition of the S-matrix is unacceptable because it introduces negative-metric states which lead to negative probabilities.

We begin by defining the set of physical asymptotic states to be those states of positive metric which do not contain quanta of action-at-a-distance fields. In an action-at-a-distance electrodynamics the set of physical states includes all electron and positron states not containing « action at a distance » photons. The problem is to define a unitary S-matrix taking « in » physical states to « out » physical states.

We choose to use a variation of Bogoliubov's procedure ([17]) for defining a unitary physical S-matrix as implemented by SUDARSHAN and co-workers ([18]). The starting point is the expansion of the S-matrix in « old fashioned » perturturbation theory:

(A.1) $$S_{\beta\alpha} = \delta_{\beta\alpha} - i(2\pi)^4\delta^4(P_\beta - P_\alpha)T_{\beta\alpha}$$

with

(A.2) $$T_{\beta\alpha} = \langle\beta|H_I|\alpha\rangle + \sum_N \frac{\langle\beta|H_I|N\rangle\langle N|H_I|\alpha\rangle}{E_\alpha - E_N + i\varepsilon} + \dots,$$

([16]) D. W. SCIAMA: *Roy. Astron. Soc. Month. Not.*, **113**, 34 (1953).

([17]) N. N. BOGOLIUBOV: *Annales of Invitational Conference on High-Energy Physics*, CERN (1958).

([18]) E. C. G. SUDARSHAN: *Fields and Quanta*, **2**, 175 (1972); C. A. NELSON: Louisiana State University preprint (1972); J. L. RICHARD: *Phys. Rev. D*, **7**, 3617 (1973); C. C. CHIANG: University Göteborg preprint (1972), and references therein.

where H_I is the interaction Hamiltonian. We wish to modify eq. (A.2) so that states with «action-at-a-distance quanta» do not contribute to the absorptive part of amplitudes. This would allow unitarity to be maintained under the restriction of unitarity sums to the set of physical states. SUDARSHAN and co-workers ([18]) have shown that this can be done by taking the energy denominator factor of unphysical states in principal value:

$$(\text{A.3})\qquad T_{\beta\alpha} = \langle\beta|H_I|\alpha\rangle + \sum_p \frac{\langle\beta|H_I|p\rangle\langle p|H_I|\alpha\rangle}{E_\alpha - E_p + i\varepsilon} + \sum_u \langle\beta|H_I|u\rangle\langle u|H_I|\alpha\rangle P\frac{1}{E_\alpha - E_u} + \dots,$$

where p labels physical states and u labels unphysical states (*i.e.* those containing action-at-a-distance quanta in our case). Equation (A.3) defines the physical T-matrix in our models. It could serve as the basis for calculating S-matrix elements.

However, we would like to re-express this S-matrix in terms of covariant perturbation theory. In order to do this we shall take advantage of Weinberg's study ([19]) of the infinite-momentum frame limit of old-fashioned perturbation theory, and Chang and Ma's realization of infinite-momentum frame results by a change of variable ([20]).

Since the difference between the physical T-matrix and the unmodified T-matrix lies only in the character of the singularity in the energy denominator, the algebraic development of Weinberg, and his power counting arguments, in particular, can be taken over without change to our case. (We also will ignore possible complications due to interchanging the order of integration and the $P\to\infty$ limit. Actually we define our T-matrix to be the result of this procedure.) The result is that Weinberg's rules apply in the present case also —except that Weinberg's rule (*d*) must be modified, so that an energy denominator is taken in principal value, if the corresponding intermediate state is unphysical—*i.e.* contains action-at-a-distance quanta.

CHANG and MA showed that it was not necessary to go to the infinite-momentum frame limit in order to realize Weinberg's rules. They showed that the expression of the usual Feynman rules in terms of infinite momentum frame variables, $\eta = p^0 + p^3$ and $s = p^0 - p^3$ for each momentum p^μ, led to the same results as Weinberg's rules. This fact can be used to advantage in the case of action-at-a-distance models. S-matrix elements can be calculated in our models from the usual Feynman rules in the following way: i) for a given diagram follow the usual Feynman rules to obtain a S-matrix contribution using infinite-momentum frame variables *à la* Chang and Ma—in particular, associate Feynman propagators with action-at-a-distance particle lines; ii) evaluate the «energy» integrals by complex integration; iii) the result will be a series of terms corresponding to different intermediate states in old-fashioned perturbation theory language. Those denominators corresponding to intermediate states with principal-value quanta should be taken in principal value. Those denominators corresponding to physical intermediate states should remain unmodified.

([19]) S. WEINBERG: *Phys. Rev.*, **150**, 1313 (1966).
([20]) S. J. CHANG and S. K. MA: *Phys. Rev.*, **180**, 1506 (1969).

Let us examine the results of this procedure for some simple cases. First, consider the pole diagram for two-particle–to–two-particle scattering. If the pole corresponds to a principal-value particle the amplitude will be proportional to

$$P\frac{1}{q^2-m^2}, \tag{A.4}$$

using the above rules ([21]). This result also follows from explicit evaluation of the diagram in canonical perturbation theory, using eq. (1) of the text for $\theta=\pi/4$.

Next consider a second-order self-energy correction due to the emission and absorption of a principal-value quantum by a particle (propagating via Feynman propagators). Our modified Feynman rules require us to evaluate an integral

$$I=i\int \mathrm{d}^4k\frac{1}{k^2-m^2+i\varepsilon}\frac{1}{(p-k)^2-\mu^2+i\varepsilon}, \tag{A.5}$$

using infinite-momentum frame variables:

$$\eta'=k^0+k^3 \qquad s'=k^0-k^3 \qquad \vec{q}=(k^1,k^2) \tag{A.6}$$

and $p=(s,\eta,\vec{0})$. The result is

$$I=\frac{\pi}{2}\int \mathrm{d}^2q\int_0^\eta \frac{\mathrm{d}\eta'}{\eta'(\eta-\eta')}\frac{1}{s-(\vec{q}^{\,2}+\mu^2)/(\eta-\eta')-(\vec{q}^{\,2}+m^2)/\eta'+i\varepsilon} \tag{A.7}$$

which must be modified to

$$I'=\frac{\pi}{2}\int \mathrm{d}^2q\int_0^\eta \frac{\mathrm{d}\eta'}{\eta'(\eta-\eta')}P\frac{1}{s-(\boldsymbol{q}^2+\mu^2)/(\eta-\eta')-(\boldsymbol{q}^2+m^2)/\eta'}. \tag{A.8}$$

In this case we see that $2I'=I+I^*$. The propagation of the particle which would ordinarily propagate via Feynman propagators is manifestly different in states where virtual action-at-a-distance quanta are present. This is necessary in order to define a unitary physical S-matrix. Thus the physical S-matrix only agrees with the canonically defined S-matrix for one–principal-value particle intermediate states and states without principal value particles.

If we consider a non-Abelian action-at-a-distance gauge theory coupled to an external classical source, it is possible to define the physical S-matrix in

---

([21]) Compare to eqs. (5) and (6) of ref. ([9]).

a different manner from the above. Consider the usual definition of the S-matrix and in particular the set of tree diagrams with no external principal-value particle lines. It is easy to verify that the absorptive part of any tree diagram is zero *in the physical region* since

$$\text{Abs}\left[P\frac{1}{k^2-m^2}\right]=0\,. \tag{A.9}$$

As a result the set of tree diagrams defines a unitary (gauge invariant) S-matrix. (This should be contrasted with the usual theory where taking the absorptive part of a tree diagram introduces states containing non-Abelian bosons which in turn introduce loops and Fadeev-Popov ghosts.) In the present case no Fadeev-Popov ghost loops are needed to maintain unitarity ([22]). Thus we wind up with a loopless, tree diagram model. The possibility of limiting the self-interaction of fields in this way allows us to develop a renormalizable model of quantum gravity.

Finally we note that the lack of contributions to unitarity sums from intermediate states containing principal-value particles allows us to avoid the introduction of Fadeev-Popov ghosts in *all* non-Abelian action-at-a-distance models.

([22]) Cf. ref. ([12]) and B. W. Lee and J. Zinn-Justin: *Phys. Rev. D*, **5**, 3121 (1972).

● RIASSUNTO (*)

Si formulano teorie di gauge nel sistema di una generalizzazione della teoria quantistica dei campi. In particolare si discutono modelli di teorie di elettrodinamica e di Yang-Mills, un modello dell'interazione forte con derivate di ordine più alto e confinamento dei quark, e un modello rinormalizzabile di gravità quantistica pura con una Lagrangiana di Einstein. Nel caso dell'elettrodinamica si mostra che due modelli sono possibili: uno con predizioni che sono identiche a QED e uno che è un modello quantistico di azione a distanza dell'elettrodinamica. Nel caso delle teorie di Yang-Mills si può costruire un modello che è identico per quanto riguarda le predizioni a qualsiasi modello convenzionale o modello di azione a distanza. Nel secondo caso è possibile eliminare tutti i cappi di particelle di Yang-Mills (in tutti i gauge) in una maniera consistente con l'unitarietà. Esiste una variazione dei modelli di Yang-Mills nella nostra formulazione che ha equazioni di campo con derivate ad ordine più alto. È unitario ed ha probabilità positive. Può essere usato per costruire un modello d'interazioni forti che ha un potenziale lineare e manifesta confinamento dei quark. Infine si mostra come costruire un modello di azione a distanza della gravità quantistica pura (il cui limite classico è la dinamica della Lagrangiana di Einstein) accoppiato ad una sorgente esterna classica. Il modello è grossolanamente rinormalizzabile.

(*) *Traduzione a cura della Redazione.*

Новый подход к калибровочным теориям поля.

Резюме (*). — Мы формулируем калибовочные теории в рамках обобщения квантовой теории поля. В частности, мы обсуждаем модели электродинамики и теорий Янга-Миллса, модель сильных взаимодействий с производными высших порядков и удержанием кварков и перенормируемую модель для чистой квантовой гравитации с Лагранжианом Эйнштейна. Мы показываем, что в случае электродинамики возможны две модели: одна модель имеет предсказания, которые идентичны предсказаниям квантовой электродинамики, и другая модель, которая представляет модель элзктродинамики с квантовым действием на расстоянии. В случае теорий Янга-Миллса мы можем сконструировать модель, которая идентична по предсказаниям любой общепринятой модели или модели с квантовым действием на расстоянии. Во втором случае имеется возможность исключить все петли для частиц Янга-Миллса (во всех калибровках). Наша формулировка содержит изменение моделей Янга-Миллса, которое включает уравнения поля с производными высших порядков. Наша модель является унитарной и имеет положительные вероятности. Наша формулировка может быть использована для конструирования модели сильных взаимодействий, которая имеет линейный потенциал и обеспечивает удержание кварков. Мы показываем, как сконструировать модель действия на расстоянии для квантовой гравитации, связанной с внешним классическим источником. Предложенная модель является перенормируемой.

(*) *Переведено редакцией.*

Direttore responsabile: CARLO CASTAGNOLI

Stampato in Bologna dalla Tipografia Compositori coi tipi della Tipografia Monograf
Questo fascicolo è stato licenziato dai torchi il 17-I-1979

REFERENCES

Akhiezer, N. I., Frink, A. H. (tr), 1962, *The Calculus of Variations* (Blaisdell Publishing, New York, 1962).

Bjorken, J. D., Drell, S. D., 1964, *Relativistic Quantum Mechanics* (McGraw-Hill, New York, 1965).

Bjorken, J. D., Drell, S. D., 1965, *Relativistic Quantum Fields* (McGraw-Hill, New York, 1965).

Blaha, S., 1998, *Cosmos and Consciousness* (Pingree-Hill Publishing, Auburn, NH, 1998).

_______, 2002, *A Finite Unified Quantum Field Theory of the Elementary Particle Standard Model and Quantum Gravity Based on New Quantum Dimensions™ & a New Paradigm in the Calculus of Variations* (Pingree-Hill Publishing, Auburn, NH, 2002).

_______, 2003, *A Finite Unified Quantum Field Theory of the Elementary Particle Standard Model and Quantum Gravity Based on New Quantum Dimensions™ and a New Paradigm in the Calculus of Variations* (Pingree-Hill Publishing, Auburn, NH, 2003).

_______, 2004, *Quantum Big Bang Cosmology: Complex Space-time General Relativity, Quantum Coordinates™ Dodecahedral Universe, Inflation, and New Spin 0, ½, 1 & 2 Tachyons & Imagyons* (Pingree-Hill Publishing, Auburn, NH, 2004).

______, *2005a, Quantum Theory of the Third Kind: A New Type of Divergence-free Quantum Field Theory Supporting a Unified Standard Model of Elementary Particles and Quantum Gravity based on a New Method in the Calculus of Variations* (Pingree-Hill Publishing, Auburn, NH, 2005).

______, 2005b, *The Metatheory of Physics Theories, and the Theory of Everything as a Quantum Computer Language* (Pingree-Hill Publishing, Auburn, NH, 2005).

______, 2005c, *The Equivalence of Elementary Particle Theories and Computer Languages: Quantum Computers, Turing Machines, Standard Model, Superstring Theory, and a Proof that Gödel's Theorem Implies Nature Must Be Quantum* (Pingree-Hill Publishing, Auburn, NH, 2005).

______, 2006a, *The Foundation of the Forces of Nature* (Pingree-Hill Publishing, Auburn, NH, 2006).

______, 2006b, *A Derivation of ElectroWeak Theory based on an Extension of Special Relativity; Black Hole Tachyons; & Tachyons of Any Spin.* (Pingree-Hill Publishing, Auburn, NH, 2006).

_______, 2007a, *Physics Beyond the Light Barrier: The Source of Parity Violation, Tachyons, and A Derivation of Standard Model Features* (Pingree-Hill Publishing, Auburn, NH, 2007).

______, 2007b, *The Origin of the Standard Model: The Genesis of Four Quark and Lepton Species, Parity Violation, the ElectroWeak Sector, Color SU(3), Three Visible Generations of Fermions, and One Generation of Dark Matter with Dark Energy* (Pingree-Hill Publishing, Auburn, NH, 2007).

______*, 2008a, A Direct Derivation of the Form of the Standard Model From GL(16) (Pingree-Hill Publishing, Auburn, NH, 2008).*

______, 2008b, *A Complete Derivation of the Form of the Standard Model With a New Method to Generate Particle Masses Second Edition* (Pingree-Hill Publishing, Auburn, NH, 2008)

______, 2009, *The Algebra of Thought & Reality: The Mathematical Basis for Plato's Theory of Ideas, and Reality Extended to Include A Priori Observers and Space-Time Second Edition* (Pingree-Hill Publishing, Auburn, NH, 2009).

______, 2010a, *Operator Metaphysics: A New Metaphysics Based on a New Operator Logic and a New Quantum Operator Logic that Lead to a Mathematical Basis for Plato's Theory of Ideas and Reality* (Pingree-Hill Publishing, Auburn, NH, 2010).

______, 2010b, *The Standard Model's Form Derived from Operator Logic, Superluminal Transformations and GL(16)* (Pingree-Hill Publishing, Auburn, NH, 2010).

______, 2010c, *SuperCivilizations: Civilizations as Superorganisms* (McMann-Fisher Publishing, Auburn, NH, 2010).

______, 2011a, *21st Century Natural Philosophy Of Ultimate Physical Reality* (McMann-Fisher Publishing, Auburn, NH, 2011).

______, 2011b, *All the Universe! Faster Than Light Tachyon Quark Starships & Particle Accelerators with the LHC as a Prototype Starship Drive Scientific Edition* (Pingree-Hill Publishing, Auburn, NH, 2011).

______, 2011c, *From Asynchronous Logic to The Standard Model to Superflight to the Stars* (Blaha Research, Auburn, NH, 2011).

______, 2012a, *From Asynchronous Logic to The Standard Model to Superflight to the Stars volume 2: Superluminal CP and CPT, U(4) Complex General Relativity and The Standard Model, Complex Vierbein General Relativity, Kinetic Theory, Thermodynamics* (Blaha Research, Auburn, NH, 2012).

______, 2012b, *Standard Model Symmetries, And Four And Sixteen Dimension Complex Relativity; The Origin Of Higgs Mass Terms* (Blaha Reasearch, Auburn, NH, 2012).

______, 2013a, *Multi-Stage Space Guns, Micro-Pulse Nuclear Rockets, and Faster-Than-Light Quark-Gluon Ion Drive Starships* (Blaha Research, Auburn, NH, 2013).

______, 2013b, *The Bridge to Dark Matter; A New Sister Universe; Dark Energy; Inflatons; Quantum Big Bang; Superluminal Physics; An Extended Standard Model Based on Geometry* (Blaha Reasearch, Auburn, NH, 2013).

______, 2014a, *Universes and Megaverses: From a New Standard Model to a Physical Megaverse; The Big Bang; Our Sister Universe's Wormhole; Origin of the Cosmological Constant, Spatial Asymmetry of the Universe, and its Web of Galaxies; A Baryonic Field between Universes and Particles; Megaverse Extended Wheeler-DeWitt Equation* (Blaha Reasearch, Auburn, NH, 2014).

______, 2014b, *All the Megaverse! Starships Exploring the Endless Universes of the Cosmos Using the Baryonic Force* (Blaha Research, Auburn, NH, 2014).

______, 2014c, *All the Megaverse! II Between Megaverse Universes: Quantum Entanglement Explained by the Megaverse Coherent Baryonic Radiation Devices – PHASERs Neutron Star Megaverse Slingshot Dynamics Spiritual and UFO Events, and the Megaverse Microscopic Entry into the Megaverse* (Blaha Research, Auburn, NH, 2014).

______, 2015a, *PHYSICS IS LOGIC PAINTED ON THE VOID: Origin of Bare Masses and The Standard Model in Logic, U(4) Origin of the Generations, Normal and Dark Baryonic Forces, Dark Matter, Dark Energy, The Big Bang, Complex General Relativity, A Megaverse of Universe Particles* (Blaha Research, Auburn, NH, 2015).

______, 2015b, *PHYSICS IS LOGIC Part II: The Theory of Everything, The Megaverse Theory of Everything, U(4)⊗U(4) Grand Unified Theory (GUT), Inertial Mass = Gravitational Mass, Unified Extended Standard Model and a New Complex General Relativity with Higgs Particles, Generation Group Higgs Particles* (Blaha Research, Auburn, NH, 2015).

______, 2015c, *The Origin of Higgs ("God") Particles and the Higgs Mechanism: Physics is Logic III, Beyond Higgs – A Revamped Theory With a Local Arrow of Time, The Theory of Everything Enhanced, Why Inertial Frames are Special, Universes of the Mind* (Blaha Research, Auburn, NH, 2015).

______, 2015d, *The Origin of the Eight Coupling Constants of The Theory of Everything: U(8) Grand Unified Theory of Everything (GUTE), S^8 Coupling Constant Symmetry, Space-Time Dependent Coupling Constants, Big Bang Vacuum Coupling Constants, Physics is Logic IV* (Blaha Research, Auburn, NH, 2015).

______, 2016a, *New Types of Dark Matter, Big Bang Equipartition, and A New U(4) Symmetry in the Theory of Everything: Equipartition Principle for Fermions, Matter is 83.33% Dark, Penetrating the Veil of the Big Bang, Explicit QFT Quark Confinement and Charmonium, Physics is Logic V* (Blaha Research, Auburn, NH, 2016).

______, 2016b, *The Periodic Table of the 192 Quarks and Leptons in The Theory of Everything: The U(4) Layer Group, Physics is Logic VI* (Blaha Research, Auburn, NH, 2016).

______, 2016c, *New Boson Quantum Field Theory, Dark Matter Dynamics, Dark Matter Fermion Layer Mixing, Genesis of Higgs Particles, New Layer Higgs Masses, Higgs Coupling Constants, Non-Abelian Higgs Gauge Fields, Physics is Logic VII* (Blaha Research, Auburn, NH, 2016).

______, 2016d, *Unification of the Strong Interactions and Gravitation: Quark Confinement Linked to Modified Short-Distance Gravity; Physics is Logic VIII* (Blaha Research, Auburn, NH, 2016).

______, 2016e, *MoND: Unification of the Strong Interactions and Gravitation II, Quark Confinement Linked to Large-Scale Gravity, Physics is Logic IX* (Blaha Research, Auburn, NH, 2016).

______, 2016f, *CQMechanics: A Unification of Quantum & Classical Mechanics, Quantum/Semi-Classical Entanglement, Quantum/Classical Path Integrals, Quantum/Classical Chaos* (Blaha Research, Auburn, NH, 2016).

______, 2016g, *GEMS: Unified Gravity, ElectroMagnetic and Strong Interactions: Manifest Quark Confinement, A Solution for the Proton Spin Puzzle, Modified Gravity on the Galactic Scale* (Pingree Hill Publishing, Auburn, NH, 2016).

______, 2016h, *Unification of the Seven Boson Interactions based on the Riemann-Christoffel Curvature Tensor* (Pingree Hill Publishing, Auburn, NH, 2016).

______, 2017a, *Unification of the Eleven Boson Interactions based on 'Rotations of Interactions'* (Pingree Hill Publishing, Auburn, NH, 2017).

______, 2017b, *The Origin of Fermions and Bosons, and Their Unification* (Pingree Hill Publishing, Auburn, NH, 2017).

______, 2017c, *Megaverse: The Universe of Universes* (Pingree Hill Publishing, Auburn, NH, 2017).

______, 2017d, *SuperSymmetry and the Unified SuperStandard Model* (Pingree Hill Publishing, Auburn, NH, 2017).

Eddington, A. S., 1952, *The Mathematical Theory of Relativity* (Cambridge University Press, Cambridge, U.K., 1952).

Fant, Karl M., 2005, *Logically Determined Design: Clockless System Design With NULL Convention Logic* (John Wiley and Sons, Hoboken, NJ, 2005).

Feinberg, G. and Shapiro, R., 1980, *Life Beyond Earth: The Intelligent Earthlings Guide to Life in the Universe* (William Morrow and Company, New York, 1980).

Gelfand, I. M., Fomin, S. V., Silverman, R. A. (tr), 2000, *Calculus of Variations* (Dover Publications, Mineola, NY, 2000).

Giaquinta, M., Modica, G., Souchek, J., 1998, *Cartesian Coordinates in the Calculus of Variations* Volumes I and II (Springer-Verlag, New York, 1998).

Giaquinta, M., Hildebrandt, S., 1996, *Calculus of Variations* Volumes I and II (Springer-Verlag, New York, 1996).

Gradshteyn, I. S. and Ryzhik, I. M., 1965, *Table of Integrals, Series, and Products* (Academic Press, New York, 1965).

Heitler, W., 1954, *The Quantum Theory of Radiation* (Claendon Press, Oxford, UK, 1954).

Huang, Kerson, 1992, *Quarks, Leptons & Gauge Fields 2nd Edition* (World Scientific Publishing Company, Singapore, 1992).

Jost, J., Li-Jost, X., 1998, *Calculus of Variations* (Cambridge University Press, New York, 1998).

Kaku, Michio, 1993, *Quantum Field Theory*, (Oxford University Press, New York, 1993).

Kirk, G. S. and Raven, J. E., 1962, *The Presocratic Philosophers* (Cambridge University Press, New York, 1962).

Landau, L. D. and Lifshitz, E. M., 1987, *Fluid Mechanics 2nd Edition*, (Pergamon Press, Elmsford, NY, 1987).

Misner, C. W., Thorne, K. S., and Wheeler, J. A., 1973, *Gravitation* (W. H. Freeman, New York, 1973).

Rescher, N., 1967, *The Philosophy of Leibniz* (Prentice-Hall, Englewood Cliffs, NJ, 1967).

Sagan, H., 1993, *Introduction to the Calculus of Variations* (Dover Publications, Mineola, NY, 1993).

Sakurai, J. J., 1964, *Invariance Principles and Elementary Particles* (Princeton University Press, Princeton, NJ, 1964).

Streater, R. F. and Wightman, A. S., 2000, *PCT, Spin, Statistics, and All That* (Princeton University Press, Princeton, NJ 2000).

Weinberg, S., 1972, *Gravitation and Cosmology* (John Wiley and Sons, New York, 1972).

Weinberg, S., 1995, *The Quantum Theory of Fields Volume I* (Cambridge University Press, New York, 1995).

Weinberg, S., 2000, *The Quantum Theory of Fields Volume III Supersymmetry* (Cambridge University Press, New York, 2000).

Weyl, H., 1950, *Space, Time, Matter* (Dover, New York, 1950).

Weyl, H., (Tr. S. Pollard et al), 1987, *The Continuum* (Dover Publications, New York, 1987).

INDEX

About the Author

Stephen Blaha is a well known Physicist and Man of Letters with interests in Science, Society and civilization, the Arts, and Technology. He had an Alfred P. Sloan Foundation scholarship in college. He received his Ph.D. in Physics from Rockefeller University. He has served on the faculties of several major universities. He was also a Member of the Technical Staff at Bell Laboratories, a manager at the Boston Globe Newspaper, a Director at Wang Laboratories, and President of Blaha Software Inc and of Janus Associates Inc. (NH).

Among other achievements he was a co-discoverer of the "r potential" for heavy quark binding developing the first (and still the only demonstrable) non-abelian gauge theory with an "r" potential; first suggested the existence of topological structures in superfluid He-3; first proposed Yang-Mills theories would appear in condensed matter phenomena with non-scalar order parameters; first developed a grammar-based formalism for quantum computers and applied it to elementary particle theories; first developed a new form of quantum field theory without divergences (thus solving a major 60 year old problem that enabled a unified theory of the Standard Model and Quantum Gravity without divergences to be developed); first developed a formulation of complex General Relativity based on analytic continuation from real space-time; first developed a generalized non-homogeneous Robertson-Walker metric that enabled a quantum theory of the Big Bang to be developed without singularities at t = 0; first generalized Cauchy's theorem and Gauss' theorem to complex, curved multi-dimensional spaces; received Honorable Mention in the Gravity Research Foundation Essay Competition in 1978; first developed a physically acceptable theory of faster-than-light particles; first derived a composition of extrema method in the Calculus of Variations; first quantitatively suggested that inflationary periods in the history of the universe were not needed; first proved Gödel's Theorem implies Nature must be quantum; provided a new alternative to the Higgs Mechanism, and Higgs particles, to generate masses; first showed how to resolve logical paradoxes including Gödel's Undecidability Theorem by developing Operator Logic and Quantum Operator Logic; first developed a quantitative harmonic oscillator-like model of the life cycle, and interactions, of civilizations; first showed how equations describing superorganisms also apply to civilizations. A recent book shows his theory applies successfully to the past 14 years of history and to *new* archaeological data on Andean and Mayan civilizations as well as Early Anatolian and Egyptian civilizations.

He first developed an axiomatic derivation of the forms of The Standard Model from geometry – space-time properties – The Extended Standard Model. It has a Dark Matter sector that approximates the ElectroWeak sector with Dark doublets and Dark gauge interactions. It also uses quantum coordinates to remove infinities that crop up in most interacting quantum field theories and additionally to remove the infinities that appear in the Big Bang and generate an inflationary growth of the universe. The Extended Standard Model has an ultra-high energy

GUT (Grand Unified Theory) limit with a U(4)⊗U(4) symmetry; and can be united with gravitation to form a Theory of Everything. (See *Physics is Logic Part II.*)

Blaha has had a major impact on a succession of elementary particle theories: his Ph.D. thesis (1970), and papers, showed that quantum field theory calculations to all orders in ladder approximations could not give scaling deep inelastic electron-nucleon scattering. He later showed the eigenvalue equation for the fine structure constant α in Johnson-Baker-Willey QED had a zero at $\alpha = 1$ not 1/137 by solving the Schwinger-Dyson equations to all orders in an approximation that agreed with exact results to 4th order in α thus ending interest in this theory. In 1979 at Prof. Ken Johnson's (MIT) suggestion he calculated the proton-neutron mass difference in the MIT bag model and found the result had the wrong sign reducing interest in the bag model. These results all appear in Physical Review papers. In the 2000's he repeatedly pointed out the shortcomings of SuperString theory and showed that The Standard Model's form could be derived from space-time geometry by an extension of Lorentz transformations to faster than light transformations. This deeper space-time basis greatly increases the possibility that it is part of THE fundamental theory.Recently, Blaha showed that the Weak interactions differed significantly from the Strong, electromagnetic and gravitation interactions in important respects while these interactions had similar features, and suggested that ElectroWeak theory, which is essentially a glued union of the Weak interactions and Electromagnetism, possibly modulo unknown Higgs particle features, be replaced by a unified theory of the other interactions combined with a stand-alone Weak interaction theory. Blaha also showed that, if Charmonium calculations are taken seriously, the Strong interaction coupling constant is only a factor of five larger than the electromagnetic coupling constant, and thus Strong interaction perturbation theory would make sense and yield physically meaningful results.

In graduate school (1965-71) he wrote substantial papers in elementary particles and group theory: The Inelastic E- P Structure Functions in a Gluon Model. Phys. Lett. B40:501-502,1972; Deep-Inelastic E-P Structure Functions In A Ladder Model With Spin 1/2 Nucleons, Phys.Rev. D3:510-523,1971; Continuum Contributions To The Pion Radius, Phys. Rev. 178:2167-2169,1969; Character Analysis of U(N) and SU(N), J. Math. Phys. 10, 2156 (1969); and The Calculation of the Irreducible Characters of the Symmetric Group in Terms of the Compound Characters, (Published as Blaha's Lemma in D. E. Knuth's book: *The Art of Computer Programming Vols. 1 – 4*).

In the early 1980's Blaha was also a pioneer in the development of UNIX for financial, scientific and Internet applications: benchmarked UNIX versions showing that block size was critical for UNIX performance, developing financial modeling software, starting database benchmarking comparison studies, developing Internet-like UNIX networking (1982) and developing a hybrid shell programming technique (1982) that was a precursor to the PERL programming language. He was also the manager of the AT&T ten-year future products development database. His work helped lead to commercial UNIX on computers such as Sun Micros, IBM AIX minis, and Apple computers.

In the 1980's he pioneered the development of PC Desktop Publishing on laser printers. and was nominated for three "Awards for Technical Excellence" in 1987 by PC Magazine for PC software products that he designed and developed.

Recently he has developed a theory of Megaverses – actual universes of which our universe is one – with quantum particle-like properties based on the Wheeler-DeWitt equation of Quantum Gravity. He has developed a theory of a baryonic force, which had been conjectured many years ago, and estimated the strength of the force based on discrepancies in measurements of the gravitational constant G. This force, operative in 15-dimensinal space, can be used to escape from our universe in "uniships" which are the equivalent of the faster-than-light starships proposed in the author's earlier books. Thus travel to other universes, as well as to other stars is possible.

Blaha also considered the complexified Wheeler-DeWitt equation and showed that its limitation to real-valued coordinates and metrics generated a Cosmological Constant in the Einstein equations.

The author has also recently written a series of books on the serious problems of the United States and their solution as well as a book on the decline of Mankind that will follow from current social and genetic trends in Mankind.

In the past twelve years Dr. Blaha has written over 40 books on a wide range of topics. Some recent major works are: *From Asynchronous Logic to The Standard Model to Superflight to the Stars*, *All the Universe!*, *SuperCivilizations: Civilizations as Superorganisms*, *America's Future: an Islamic Surge, ISIS, al Qaeda, World Epidemics, Ukraine, Russia-China Pact, US Leadership Crisis,The Rises and Falls of Man – Destiny – 3000 AD: New Support for a Superorganism MACRO-THEORY of CIVILIZATIONS From CURRENT WORLD TRENDS and NEW Peruvian, Pre-Mayan, Mayan, Anatolian, and Early Egyptian Data, with a Projection to 3000 AD,* and *Mankind in Decline: Genetic Disasters, Human-Animal Hybrids, Overpopulation, Pollution, Global Warming, Food and Water Shortages, Desertification, Poverty, Rising Violence, Genocide, Epidemics, Wars, Leadership Failure.*

He has taught approximately 4,000 students in undergraduate, graduate, and postgraduate corporate education courses primarily in major universities, and large companies and government agencies.

The above paragraphs summarize much of his work over the past fifty years. This work is fully documented. He continues to engage in research and writing at Blaha Research.

www.ingramcontent.com/pod-product-compliance
Lightning Source LLC
Chambersburg PA
CBHW082006190326
41458CB00010B/3091